THE AVIATOR'S GUIDE TO
MODERN NAVIGATION

DONALD J. CLAUSING

TAB BOOKS
Blue Ridge Summit, PA

This publication contains chart illustrations that have been reproduced with the permission of, and are copyrighted by, Jeppesen Sanderson, Inc. They are for purposes of illustration only, and in most cases have been enlarged or reduced in size for additional clarity. They are not to be used for navigation. Chart information was current at the time of writing, but is subject to change.

To CPT John R. ''Robbie'' Peacock II, USMC, Aviator.
Darmouth College, Class of 1968.
Missing in Action over North Vietnam, October 12, 1972.

FIRST EDITION
FIFTH PRINTING

© 1987 by **TAB Books**.
TAB Books is a division of McGraw-Hill, Inc.

Printed in the United States of America. All rights reserved. The publisher takes no responsibility for the use of any of the materials or methods described in this book, nor for the products thereof.

Library of Congress Cataloging-in-Publication Data

Clausing, Donald J.
     The aviator's guide to modern navigation.

     Includes index.
     1. Navigation (Aeronautics)   I. Title.
TL586.C534   1987   629.132′51   86-23194
ISBN 0-8306-0208-9
ISBN 0-8306-2408-2 (pbk.)

TAB Books offers software for sale. For information and a catalog, please contact
TAB Software Department, Blue Ridge Summit, PA 17294-0850.

Questions regarding the content of this book should be addressed to:

Reader Inquiry Branch
TAB Books
Blue Ridge Summit, PA 17294-0850

Front cover: large background photograph courtesy of Piper Aircraft Corporation. Small inset photograph courtesy of Sperry Corporation.

Table of Contents

NAUTICAL

Acknowledgments

NAUTICAL |⌐ 0 | | | | | 10 | | | | | 20 | | | | | 30 | | | | | 40 | | | | | 50 | | | | | 60 | | | | | 70 | | | | | 80 ⌐|

Many people were very helpful in providing background information, illustrations, and advice in the preparation of this manuscript. While it is impossible to list all who contributed, a list of some of those who were especially helpful would include Ed Phelps (AMCA International), Roger Hart (Canadian Marconi), Mari Schneider (Collins Avionics), Jerry Sciez (Delco Systems), Norman Weil and William Williams (Federal Aviation Administration), Scott Mackey (Global Systems), James Terpstra (Jeppesen), Pat Millard (Narco), Pat Bloodworth (Offshore Navigation), Janice Baker (Sperry), Joan Drisdale (Tracor), Dick Thomas (II Morrow), and Ken Llewellyn (Universal Navigation).

I would like to specifically thank Richard Neumann, Ph.D., for his help in areas of math and science, Francis Johnson for his diligent proofreading and copy editing, and David Maxwell for his excellent darkroom work. In addition, I want to thank Capt. Nathan Lake, my colleague in the AMCA International Flight Department and a trained Navy navigator, for his expert help in pointing out awkward and misleading phraseology, technical errors and misstatements, and tactful advice. All of these individuals were most helpful in the preparation of this manuscript; however, the errors that remain are the responsibility of the author alone.

Introduction

NAUTICAL 0   10   20   30   40   50   60   70   80

Many books have been written on the general principles of air navigation, but most of these books have focused on the basic techniques of visual navigation—techniques, such as pilotage and dead reckoning, that are typically taught to student pilots. A handful of books have been written on specialized navigational subjects such as RNAV, the use of LORAN-C, and over-water navigation, but there are no books currently in print which deal thoroughly with the entire subject of modern air navigation—dead reckoning, VOR, DME, NDB, RNAV, RADAR, LORAN-C, OMEGA, INS, NAVSTAR/GPS, EFIS, and navigation management. The purpose of this book is to fill that gap.

This book looks at every aspect of modern instrument navigation, from the basics of dead reckoning and VOR navigation, to current instrument airway and approach navigation and advanced techniques of area navigation. All long range nav systems are covered in detail, including LORAN-C, the fastest growing area of long range navigation at the moment. Specialized subjects not normally covered in basic navigation texts are included as well: over-water navigation, non-radar navigation, terrain mapping, electronic flight information systems, and navigation management systems—the "Navigator's Navigator." This book also looks ahead to the changes that can be expected as satellite-based systems of area navigation become operational in the near future. Every pilot, professional or non-professional, and regardless of his level of ability or experience, should be able to find areas of interest in this book, and all should be able to benefit from the review of principles and techniques previously encountered.

In terms of prior knowledge, this book assumes nothing except an understanding of the basics of aviation—a student pilot at the cross-country stage should have no trouble with any of the principles or explanations. (A familiarization with high school level algebra, geometry, and physics is helpful in understanding some of the navigational principles involved, but is not essential to an understanding of the practical applications.) At the other end of the experience scale, pilots of multiengine and turbine aircraft, including pilots of corporate, commuter, and airline equipment, should find many items of interest and practical application, especially in the areas of long range navigational systems, electronic displays, and navigation management.

My goal in writing this book has been to make all aspects of navigation understandable at the student pilot level, while still being thorough enough to be useful to more experienced pilots. The book can be read straight through, or selectively, as a reference. The beginning pilot may want to build on the basics covered in the early chapters before moving on to the more advanced subjects covered in later chapters; the more experienced pilot may want to go straight to the subjects that most interest him.

Whenever a subject area or word which may be new to the inexperienced pilot is mentioned in passing or by way of example, the location of a more detailed explanation later on in the book is provided in parentheses. (For instance, "For a more detailed discussion of INS systems, see Chapter 11.") The more experienced reader should feel free to skip about—if he finds he has gotten ahead of himself, or needs to refresh his memory about a certain subject, references to subjects previously covered are also provided. A glossary is provided at the end of the book of all technical terms, and a list of all abbreviations used in the book precedes the glossary. (All terms listed in the glossary are printed in italics when first encountered, and all abbreviations are spelled out when first encountered.)

My aim has been to create both a complete guide to modern air navigation, as well as a practical guide to navigation. Toward this end, a list of additional sources of information and flight planning assistance is included in Appendix A. A list of chart sources is provided in Appendix B, and a complete list of navigation symbols is included in Appendix C. Readers with suggestions to improve or expand future editions are encouraged to send their comments to me in care of TAB BOOKS Inc., P.O. Box 40, Blue Ridge Summit, PA 17214.

A note on pronouns: as the father of two girls, one of whom has already expressed an interest in learning to fly, I am very aware of the problem inherent in always referring to pilots as "he." Many pilots are women, and to refer to all pilots as "he" seems wrong. On the other hand, the constant use of "he or she" is repetitious and wordy, and switching from "he" to "she" at regular intervals can be confusing. I don't have an answer for this, so I have done what has always been done, and that is I refer to pilots as "he" when I mean "he or she." I wish there were a better solution.

Chapter 1

Pilotage and Dead Reckoning

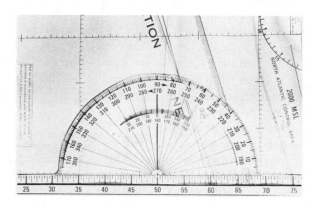

Navigation is a difficult term to define, but a perfectly accurate definition is hardly necessary—we all know, intuitively, what "navigation" means—it means knowing where you are, where you want to go, and having a good idea of how much time and fuel it will take to get there. Navigation means finding your way.

The two most fundamental methods of finding your way in an airplane are *pilotage*—the identification of present position and direction of flight by visual contact with the terrain—and *dead reckoning*—the application of fundamental laws of motion to estimate present position and predict future position. The two are so basic that they are generally thought of together by most pilots as if "pilotage and dead reckoning" was a single word. In fact, the two are so interdependent that the two methods are, for all practical purposes, essentially one method. (While it is possible to navigate exclusively with pilotage, making your way from one terrain feature to another, it is risky. On the other hand, only a fool would fail to look out the window to supplement his dead reckoning if he could.) We assume in this book that pilotage will be used whenever it is possible to use it, and that when we talk about any individual navigational method or system, we also mean "as supplemented by pilotage whenever possible."

In the purest sense, dead reckoning means the determination of position by *advancing* a previous position—to "advance" a previous position means to move it forward to its next estimated position, based on estimates of aircraft speed and direction. Each position so advanced is then used in turn to estimate the next position until, eventually, the desired destination has been reached. This is the technical meaning of dead reckoning. In a more general and less technical sense, dead reckoning means the calculation of all basic flight parameters necessary to safely navigate from point A to point B: present position, distance to go, ground speed, aircraft heading, estimated fuel flow, total fuel required, estimated time en route, and estimated time of arrival.

Dead reckoning is both the most basic and the most sophisticated form of navigation there is. Dead reckoning was used thousands of years ago by the Phoenicians to navigate at night and out of sight of land, and dead reckoning is used today to put space vehicles into orbit. Dead reckoning is, in fact, the basis of all navigation. It can be used by itself as a primary source of navigational guidance; it can be used to supplement and provide redundancy for other electronic navigational aids; and, it can form the basis for a completely self-contained system of navigation that is used not only in air navigation, but also in submarine, space, and missile navigation.

Since dead reckoning is so fundamental to all navigation, most books on air navigation begin with dead reckoning, as this book does as well; however, most books on navigation for general aviation limit their treatment of dead reckoning to its application under Visual Flight Rules (VFR). There is nothing wrong with this, but dead reckoning is much more important, and also much more complex, than a simple form of visual navigation. Dead reckoning is an important component of navigation under Instrument Flight Rules (IFR) as well—the military, for instance, has always treated dead reckoning as an important method of air navigation for all its pilots. (The military has to be prepared to operate under tactical con-

ditions where the use of electronic navigational aids could lead to detection, or where those aids were unavailable due to battlefield damage. Under these circumstances, dead reckoning may be the only method of navigation available.)

Tactical considerations are not factors in civil aviation, but to the extent dead reckoning forms the basis for all air navigation, regardless of the electronic navigational aids being used, and to the extent all pilots need to have an ultimate back-up navigational system, dead reckoning is still an important area of modern air navigation, and is worth studying in some detail. In fact, once experienced pilots understand that dead reckoning is actually a form of *area navigation*—operation along random routes without the need to overfly ground based facilities—they generally look at dead reckoning in an entirely new light.

PRINCIPLES OF DEAD RECKONING

This section on the principles of dead reckoning is not meant to be a substitute for primary flight instruction—if you have not reached the cross-country

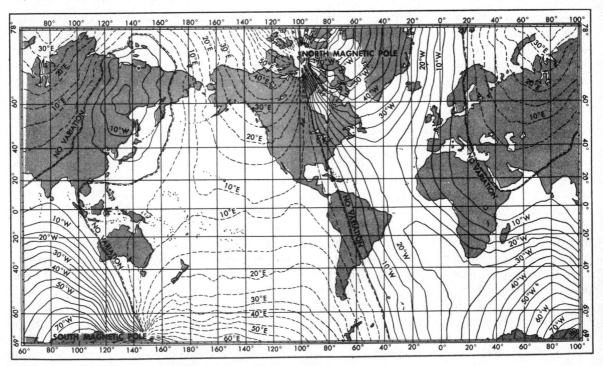

Fig. 1-1. Magnetic variation is not a constant, but varies with location.

level yet as a student pilot, you may want to refer to that section in your flight training manual first—nor is it meant to be a detailed, step-by-step instruction manual in flight planning for student pilots. Rather, what follows is a review of the principles of dead reckoning, with examples of the ways in which dead reckoning can be used both in visual and in instrument flight planning and navigation.

True North—Magnetic North

Because Earth is a sphere, and because every point on a sphere is the same as another, the only way to describe the location of something on Earth is in relation to some other point. This requires a starting point, and while (in theory) any point will do, in practice, both *True North* and *Magnetic North* are the most commonly used references. (True North is located at one end of the axis about which Earth rotates, while Magnetic North is a point near True North that is used as a convenient and practical substitute.)

True North is fine as a chart reference, but it is difficult to locate. Magnetic North is fairly easy to find, but it is somewhat unstable and it varies in its relationship to True North. Neither reference is perfect, but together they form a basis for describing location and direction between any two points on Earth, and pilots must be able to use each with equal ease.

Pilots must also be aware of the need for consistency in their use: you cannot compare one direction (say, wind direction) in terms of True North, and another direction (heading, for instance) in terms of Magnetic North, without introducing the possibility for substantial error—as much as 70 degrees in some parts of the world.

Fortunately, it is a fairly simple matter to convert from True North to Magnetic North, and vice versa. The difference between True North and Magnetic North is called *Magnetic Variation*. (See Fig. 1-1.) The amount of variation for any given location is marked on all aeronautical charts in terms of the amount and direction Magnetic North varies from True North. (See Fig. 1-2.) Thus, 12 degrees East variation means that Magnetic North is actually 12 degrees to the east of True North for that particular location. Rather than have to analyze the relationship

between True and Magnetic North each time a conversion has to be made, pilots frequently rely on the saying "East is least and West is best", meaning, easterly variation is a negative quantity, and westerly variation is a positive quantity. The necessary correction can then be made by adding the variation to True North:

$$\text{True North} + \text{Variation} = \text{Magnetic North}$$

With this formula, it is a fairly simple matter to convert from true to magnetic and from magnetic to true. For instance, to convert a True direction of 030 to Magnetic with a variation of 9 degrees East, 9 degrees East becomes negative 9 ("East is least"). Filling in 030 for TRUE NORTH and −9 for VARIATION in the formula gives us :

$$030 + (-9) = \text{Magnetic North}$$
$$\text{Magnetic North} = 021$$

Given the Magnetic direction of 021, and plugging 021 and −9 into the formula, we have:

$$\text{True North} + (-9) = 021$$
$$\text{True North} = 021 - (-9)$$
$$\text{True North} = 021 + 9$$
$$\text{True North} = 030$$

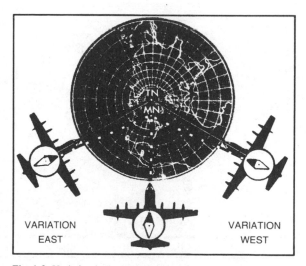

VARIATION EAST

VARIATION WEST

Fig. 1-2. Variation is the angle between True North and Magnetic North.

Courses

A course is a line drawn on a map from point A to point B. The course direction is a measure of the angle that course makes with a line referenced either to True or to Magnetic North. When a line is drawn on a chart and measured against True North grid lines (see Fig. 1-3), the result is True Course. When that course has been converted to Magnetic North, it is a Magnetic Course.

Great Circle Courses. A Great Circle is the circle formed when you slice a sphere through the center—in other words, when you cut it in half. (Logically enough, a Small Circle is formed when you cut a slice, but without going through the center.) The shortest distance between any two points on a sphere is described by the Great Circle that slices through those two points. It can be determined by drawing a string tightly from one point on a sphere (or globe) to another, or with the use of specialized charts (Gno-

monic Projections) developed specifically for this purpose.

If you draw a string tightly from one point on a globe to another, you will see that the string crosses the True North lines at slightly different angles as the Great Circle course crosses over those lines. (See Fig. 1-4. True North lines are also called longitudinal lines, lines of longitude, or meridians—they all mean the same thing.) Since Great Circle courses are described in terms of the angle they make with the True North lines, this means that a Great Circle route must be described in terms of continuously changing numbers of degrees True North; therefore, the shortest distance between two points on a sphere is not a single direction or angle, but a constantly changing one. (Actually, a Great Circle course is a constant direction, but because of the way we measure course angles it appears to change.)

Rhumb Line Course. The problem inherent in try-

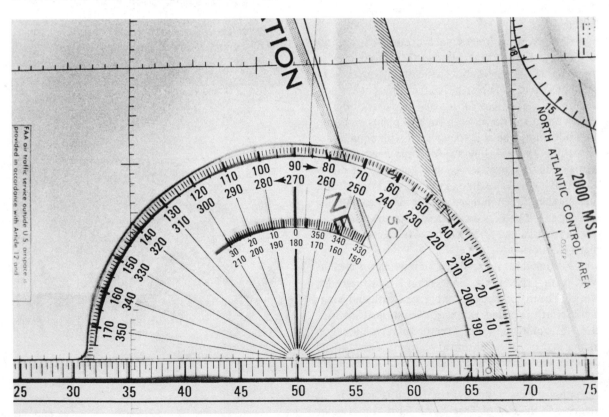

Fig. 1-3. When a line is drawn on a chart and measured against True North grid lines, the result is True Course.

Fig. 1-4. A Great Circle route crosses each meridian at a slightly different angle from the one before it.

ing to fly a course that is constantly changing is avoided by simply drawing a straight line on the chart and ignoring the error involved, and this kind of course is called a Rhumb Line course. Over short distances, and for latitudes away from the poles (latitudes are the lines that run East and West, also called parallels), the difference in distance between a Great Circle course and a Rhumb Line course is minimal. A Rhumb Line course is much easier to fly than a Great Circle course—the heading does not have to be adjusted with changes in longitude.

Over longer distances—transatlantic for instance—the difference between a Great Circle course and a Rhumb Line course can be considerable, and cannot be ignored. The calculation of automatic Great Circle routes is built in to most long range navigational systems (such as LORAN-C and OMEGA/VLF, to be covered in later chapters), which eliminates the problem of flying a course that appears to be constantly changing in direction. Since virtually no one navigates entirely via basic dead reckoning over very long distances anymore, this distinction between Great Circle and Rhumb Line courses may appear to be academic, but area navigation is becoming more and more important, and pilots should understand the difference between Great Circle routes

and Rhumb Line courses, even if they seldom encounter situations in their day-to-day flying where the distinction is significant.

Heading and Track

Heading is the actual direction in which the aircraft is pointed, or headed. If course is the direction in which the pilot wants the aircraft to go, heading is the direction he points the aircraft to achieve that course (allowing for the effect of crosswinds). *Track* is the course actually flown, or "the course made good" as it is sometimes expressed. (Course is also sometimes described as the Desired Track.)

Heading is determined by adding or subtracting a Wind Correction Angle (WCA) to the desired course. (The calculation of wind correction angles is covered in a later section.) *Heading* is obtained when a WCA is applied to a True Course. However, in order for True Heading to be something that is usable in the aircraft, it must be converted to *Magnetic Heading,* using the True North + Variation = Magnetic North formula.

Compass Heading is Magnetic Heading corrected for any error inherent in the compass itself. This error is called *Deviation*, and represents the amount, in degrees, the compass deviates from what it would read in a perfect installation. Every compass has with it a Compass Card which indicates the amount of deviation, usually in the form of Compass Headings to fly for various Magnetic Headings. (See Fig. 1-5.) For Magnetic Headings that fall between those listed, interpolation is required.

Thus, starting with a True Course—a line drawn on a map—there is a natural progression of courses and headings, resulting finally in a Compass Heading to fly to achieve the True Course desired:

True Course + Wind Correction Angle (Left −, Right +)
= True Heading

True Heading + Variation (East −, West +) = Magnetic Heading

Magnetic Heading + Deviation (+ or −) = Compass Heading

This is frequently expressed on flight planning worksheets and flight logs as:

TC + WCA + VAR + DEV = CH.

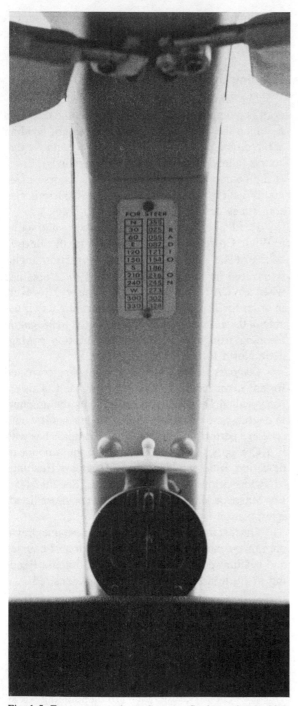

Fig. 1-5. Every compass has a Compass Card associated with it that indicates the amount of deviation for that particular installation in the form of Compass Headings to fly, or "Steer," for various Magnetic Headings.

True Airspeed

Accurate True Airspeed (TAS) information is essential to dead reckoning. Since the air itself is constantly changing both in temperature and in pressure, measuring speed through the air is not a matter of simply measuring the pressure of the ram air on a sensor. Both temperature and pressure affect the *density* of the air. Less dense air acts on an airspeed indicator like slower air, and denser air acts like faster air. The only time the indicated airspeed (IAS) and the TAS will be equal is when the air is neither more nor less dense than the standard to which the airspeed indicator is calibrated.

The internationally agreed upon standard atmosphere is a barometric pressure of 29.92 inches of Mercury at the surface, at a temperature of 15 degrees Celsius, decreasing approximately 2 degrees Celsius and one inch of Mercury for every thousand feet of altitude above Mean Sea Level (MSL). This means that at any altitude greater than sea level the air will be less dense than at sea level (everything else being equal), and the indicated airspeed will be *slower* than actual true airspeed. For any temperature *warmer* than standard (for that altitude), the airspeed will also indicate slower than True, and for any altimeter setting *less* than 29.92 inches of Mercury, the indicated airspeed will be an additional amount slower than True. (The opposite in each case is also true.) The errors can work all in one direction, or in various combinations and amounts of plus and minus. Thus, in the real world, Indicated Airspeed almost never exactly equals True Airspeed, and the exact difference is—for all practical purposes—impossible to determine except by calculation. (There is such a thing as a True Airspeed Indicator, but this is simply a standard airspeed indicator that calculates TAS internally.)

A Dead Reckoning (DR) computer, often referred to as an E-6B (its original military designation), or "whiz wheel", is used to convert IAS to TAS. A DR computer is not a computer in the modern sense of an electronic, binary number processor, but in the literal sense of something that computes. Recently, electronic calculators with built-in DR functions have come on the market, as well. An example is shown in Fig. 1-6. These specialized calcu-

Fig. 1-6. Electronic navigation computers are becoming more and more common, and are beginning to replace the traditional "E-6B" Dead Reckoning computer.

Calibrated Airspeed on the inner scale. (Calibrated Airspeed [CAS] is IAS corrected for installation error, and must be found in the aircraft performance manual; normally this error is small and IAS is used instead.) Higher performance aircraft can obtain more accurate TASs with DR computers that also correct for recovery coefficient, an error involving the compressibility of air at high speeds. A typical example of this type is shown in Fig. 1-7. Detailed procedures for converting IAS to TAS are outlined in the instruction manuals for each type.

True Airspeed is used in flight planning to estimate ground speed from forecast winds aloft data, and is used in flight to check power settings and determine tailwind components from measured ground speeds. Estimated TAS for any given altitude and power setting can be found in the aircraft performance manual. It is a key element in preflight navigational planning.

Wind Correction Angles

Wind causes an aircraft to drift from the desired course in direct proportion to the strength of the crosswind component of the wind, and in indirect proportion to the TAS of the aircraft. The tendency to drift can be negated by turning the aircraft into the wind by an amount appropriate to the drifting tendency. The amount of compensation added or subtracted to the no-wind aircraft heading is called the Wind Correction Angle; given TAS, course, wind direction, and wind velocity, the amount of wind correction can be estimated in advance with the DR computer.

There are two common types of DR computers, the slide type (see Fig. 1-8), and the circular type. The slide type is slightly more accurate and a little easier to use than the circular type, but the circular type is smaller and a little easier to carry. Slide types solve wind problems graphically; electronic navigation computers and circular types solve wind problems trigonometrically. Detailed instructions for determining WCAs are given in the manuals which accompany each type. Regardless of the type of computer, TAS, true course, wind direction, and wind velocity must be known to determine a wind correction angle.

lators are easy to use, accurate, and fast, but they are also more expensive than mechanical computers. Since the mechanical DR computer is still far more common than the electronic type, all of our examples will be based on the mechanical computer.

There are actually several different methods of converting IAS to TAS; the method used depends upon the computer, the type of aircraft, and the level of operation. Some methods are more accurate than others, and some require Mach data. The simpler solutions found on basic computers are adequate for most unpressurized, general aviation aircraft. They involve aligning a pressure altitude scale (this compensates for air pressure and altitude) with a true outside temperature scale (temperature compensation). The TAS can then be read on the outer scale over

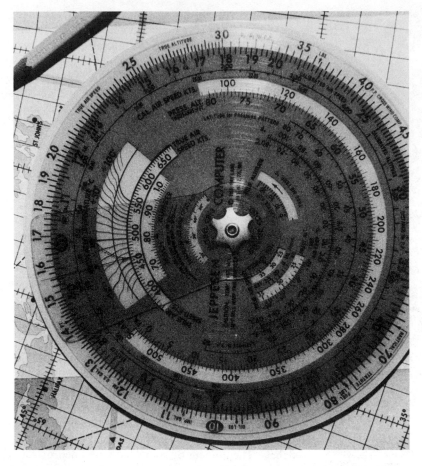

Fig. 1-7. An example of a DR computer of the type suitable for solving TAS problems for high performance aircraft.

Ground Speed

Wind, in addition to causing the aircraft to drift off course, also affects the speed with which the aircraft travels over the ground. Any given wind can be broken down into two parts, or components: a crosswind component, and a tailwind component. The crosswind component is the part that causes the aircraft to drift, and the tailwind component is the part that causes the aircraft to travel over the ground at a faster or slower rate. (A headwind is a negative tailwind in navigation. This is easier than saying "headwind or tailwind component", and it is consistent with what a headwind does: *subtracts* True Airspeed.) The tailwind component, when added to the True Airspeed, produces Ground speed (GS). En route, GS is normally more important than TAS, because GS is what determines how long it will take to reach a certain point and how much fuel will be required.

True Airspeed is more important in preflight planning, since TAS is used to estimate GS.

The wind side of the DR computer is used to determine GS as well as WCA. The instruction manual for the specific computer to be used should be consulted for this procedure; generally, only one or two additional steps are required to estimate GS once a WCA has been determined. The actual amount of the tailwind component can be determined by subtracting TAS from ground speed. For instance, a GS of 180 with a TAS of 150 would mean a tailwind component of 30 knots: $180 - 150 = 30$. (Remember, a positive value represents a tailwind, and a negative value represents a headwind.)

Actual ground speed (versus estimated ground speed) can be determined in flight given elapsed time and distance between two points. Using the slide rule side of the DR computer, distance (on the outer scale)

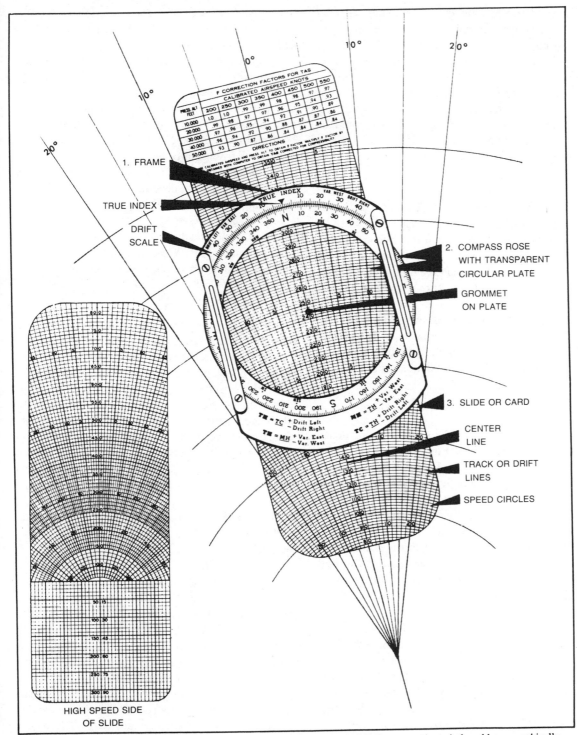

Fig. 1-8. The most common and easiest to use type of DR computer has a slide for computing wind problems graphically. Circular types solve wind problems trigonometrically.

is aligned with time (on the inner scale). Ground speed is read over the pointer. For instance, if you were to cover 33 nautical miles in 9 minutes, 33 would be set over 9, and the ground speed, 220 knots, would be read over the pointer. (The numbers on a slide rule are written in units from 10 to 100, but ''10'' can mean 0.1, 1.0, 10, 100, and so on—you have to keep the number of places straight yourself. In this case, the number indicated is 22, and the only reasonable answer is 220 knots.)

Time

Given distance and ground speed, estimated time en route is a simple calculation: distance divided by ground speed equals time. On the DR computer, ground speed goes over the pointer, and time is read on the inner scale under distance on the outer scale. Thus, an aircraft traveling 220 knots over the ground would need 30 minutes to cover 110 nautical miles: with 220 (knots ground speed) over the pointer, 30 (minutes), the answer, would be under 110 (distance). (What you are actually doing is setting up a proportion: 220 nautical miles is to 60 minutes as 110 nautical miles is to what?)

Fuel

Both fuel flow and fuel required can be determined with the DR computer in the same way ground speed and time en route are. If you know how much fuel you have burned for a given amount of time, then that becomes the basis for determining the rate of fuel flow in pounds or gallons per hour. Fuel burned, on the outer scale, goes over time on the inner scale, and fuel flow per hour is read over the pointer—the ''rate per hour'' symbol. For instance, if the aircraft has burned 300 pounds of fuel in 50 minutes, then fuel flow rate would be 360 pounds per hour (300 is to 50 as 360 is to 60).

Once fuel flow is known, fuel required for a given amount of time aloft can be determined. Thus, with a fuel flow of 360 pounds per hour (PPH), and an estimated time remaining of 1 hour and 10 minutes until over the destination (or any other point en route), 360 would go over the pointer, and fuel required—420—would be read over 70 (1 hour and 10 minutes, in minutes). The problem can also be set

up to give time remaining for a given fuel flow (endurance). For instance, with the same fuel flow, 360 PPH, but with actual fuel remaining of 600 pounds, how much time is left until the fuel is gone? Rates always go over the pointer, so with 360 over the pointer, the answer, 100 (minutes) appears under 600.

With accurate TAS, course, distance, fuel flow, and winds aloft information, the DR computer can be used to determine all the necessary information to navigate from A to B: wind correction angle, heading, ground speed, time en route, fuel flow, and fuel required. In this way, dead reckoning (supplemented by pilotage) can be used for primary navigation under VFR, and for supplemental navigation (flight planning, time and fuel estimates, and emergency heading) under IFR. We will look at both VFR and IFR applications in the next two sections.

VFR APPLICATIONS

When the weather is good, dead reckoning can be an effective method of navigation for VFR operations. Exactly what constitutes good weather is a matter of pilot judgement and not a hard-and-fast rule, and with all of the navigational aids normally available to the general aviation pilot (VOR, DME, NDB, all covered in the next few chapters) there is certainly no reason to ever risk the inaccuracies of dead reckoning when the visibility and ceiling are marginal and other methods of navigation are available. But when the visibility is good enough to provide easy visual confirmation of course and ground speed (pilotage), dead reckoning can not only be an efficient and safe method of navigation, but an enjoyable and rewarding one.

Sectional Charts (visual navigation charts to a scale of 6.86 nautical miles per inch) are normally used for VFR navigation. Straight lines connecting any two points on a Sectional Chart are Rhumb Line courses, not Great Circle routes, but for short to medium range flights the difference in distance is negligible, and as a practical matter, a Rhumb Line course is much easier to plot and fly.

To plot a Rhumb Line course, locate the departure and destination airports and draw a course line between the two with a straight edge. The course direction is measured with a protractor (or more commonly, with a combined protractor/plotter, usually shortened to "plotter"). This is done by aligning the protractor at the junction of the course line and an intersecting North/South grid line (refer back to Fig. 1-3), and reading the degrees indicated on the protractor under the course line (being sure to use the scale appropriate for the direction of flight.)

Dead reckoning for VFR is not difficult, but it does involve several steps and it produces a lot of numbers; the best way to organize the process is with a worksheet, and the best way to record the data is with a flight log. Figure 1-9 shows a typical, combined worksheet and flight log. Basic flight planning data and E-6B computations go along the top row—the worksheet section; en route data is entered under the appropriate headings below. (We will be referring to this flight log frequently as we proceed step-by-step through the planning process. It is based on a hypothetical VFR flight between Martha's Vineyard, just off Cape Cod, and Meriden Airport, near Hartford, Connecticut. A New York Sectional or a CF-19 WAC chart can be used to follow along.) As a first step, the True Course for this flight, 274

degrees, has been entered on the flight log in Fig. 1-9 under the block labeled TC.

Having drawn a course line and determined its True Direction, the next step is to break that course into en route segments based on easily recognizable terrain features, measure the distance between those segments with a plotter, and add the individual segments to determine total distance. This information goes in the flight log section, with the departure point, MVY (Martha's Vineyard), on the top row, the individual segments below that (numbered here to correspond to numbered visual fixes on the chart), and the destination, MMK (Meriden) listed as the last fix.

The distance between MVY and Fix #1 (14 nautical miles) is entered under Dist (distance) on the first line of the en route portion, then the distance between Fix #2 and Fix #3 (22 nautical miles) is entered under that, and so on for each of the segments. Then the individual segment distances are totaled, and that figure (99) is entered at the top under Total Dist. For the REM column (for distance REMaining), we start with "0" distance remaining at the destination and then work up, adding each segment distance to

E-6B COMPUTATIONS

TAS	FUEL FLOW	TC	WIND DIR	WIND KTS	WCA	TH	VAR -E +W	MH	DEV	CH
147 KTS	11.5 GPH	274	330	30	10 R	284	14 W	298	2	300

FLIGHT LOG

DEPT POINT			TOTAL DIST			TOTAL TIME		:HOURS : MINS	TIME OFF		FUEL ON BOARD
MVY			99			47		0 : 47			44

: FIX :	HDG :	DIST :	REM :	EST GS :	ACT GS:	ETE :	ATE :	ETA :	ATA :	ESTIMATED FUEL : REM :	ACTUAL: REM :
1	300	14	85	128		7				1.3 : 42.7	
2	300	22	63	128		10				2.0 : 40.8	
3	300	31	32	128		15				2.8 : 38.0	
4	300	19	13	128		9				1.7 : 36.3	
MMK	300	13	0	128		6				1.2 : 35.1	

TC + WCA + VAR + DEV = CH

Fig. 1-9. A typical VFR worksheet and flight log as it might look prior to takeoff.

the previous distance remaining until we find ourselves back at the top. (When the last distance is added to the last distance remaining, it should equal Total Dist.) We thus end up with distance information in three different forms: Total Distance; Distance between Fixes; and Distance Remaining.

Course and distance information is constant for each trip over the same route—once determined, it can be filed and retrieved whenever the trip is repeated. True Airspeed, fuel flow, and wind data however, will vary with each trip (TAS and fuel flow vary with cruising altitude, and winds vary with the weather). In our example we assume a cruising altitude of 4,000 feet (actually 4,500 feet for VFR flight in a westerly direction) and 55 percent power. (For a complete discussion of altitude and cruise power selection, see *Fly Like a Pro*, also by the author, TAB BOOKS #2378). The performance manual for our aircraft shows a TAS of 147 knots and a Fuel Flow of 11.5 gallons per hour (GPH) at 4,000 feet and 55 percent power. This information has been entered on the top row under the appropriate headings (Fig. 1-9).

Winds aloft data in the VFR range (forecast wind direction and speed for 3,000, 6,000, 9,000, 12,000 and 18,000 feet), can be obtained from any Flight Service Station (FSS) for various reporting points in the United States. For this flight, we are using the data for JFK (John F. Kennedy Airport)—the nearest reporting point in this case. On a longer trip, where the winds varied significantly in direction or strength along the route of flight, we would use the winds aloft data for the reporting point closest to each segment, computing as many different WCAs and GSs as there are reporting points. For routes that fall directly between two reporting points, and for altitudes between each of those reported, the wind direction and speed should be interpolated. In this example, we have determined (using winds for JFK, and interpolating between forecasts for 3,000 and 6,000 feet) that winds aloft are forecast to be from 330 degrees True (winds aloft are always given in terms of True North) at 30 knots. That information is entered in the worksheet section.

Using the DR computer, with 274 for the TC, 147 knots for the TAS, and winds from 330 True at 30 knots (note that all data are in consistent units of measure: True North for course and wind direction, knots for airspeed and wind speed), we obtain a wind correction angle of 10 degrees Right. This means that a heading 10 degrees to the right of course—284 degrees True—should negate the drift caused by the northwesterly wind, causing the aircraft to track along the desired course. True Heading is therefore 284 degrees, and 284 is entered under TH (Fig. 1-9).

True Heading is converted to Magnetic Heading using the familiar TH + VAR = MH formula. The amount of variation is shown on the Sectional Charts by *Isogonic Lines*—lines of constant magnetic variation (See Fig. 1-10.) Variation for this area is between 14 and 15 degrees West. Averaging the two to 14.5 degrees results in a figure that is beyond the accuracy of the compass, so, in this case, we elect to use the isogonic line closest to the destination: 14 degrees West. (An alternate and more accurate method is to use 15 degrees West for the first half of the flight, and 14 degrees West for the second half, but for a short flight like this in good visual conditions, this is unnecessary.) Plugging 14 (positive 14—''West is best,'') and 284 into the formula gives us 284 + 14 = 298: a Magnetic Heading of 298 degrees. This is the estimated magnetic heading to fly to achieve a True Course of 274 degrees, allowing for our flight planned TAS, forecast wind direction and speed, and magnetic variation.

Unfortunately, as noted earlier, compasses themselves have errors; that is, they are not perfectly accurate in all directions due to magnetic disturbances in the aircraft that cannot be fully compensated for. The amount of deviation is a function of aircraft heading, and is recorded on the compass card (Fig. 1-5). The closest direction to 298 under the ''For'' column is 300, and the compass heading to fly for 300 is 302, or plus 2. Adding 2 to 298 gives us a Compass Heading of 300 degrees. This heading, 300, is the actual heading to fly, using the aircraft magnetic compass, in order to achieve a True Course of 274 degrees, allowing for flight planned TAS, forecast wind direction and speed, magnetic variation, and finally, magnetic deviation.

This completes the dead reckoning estimate of aircraft heading. The only remaining estimates are

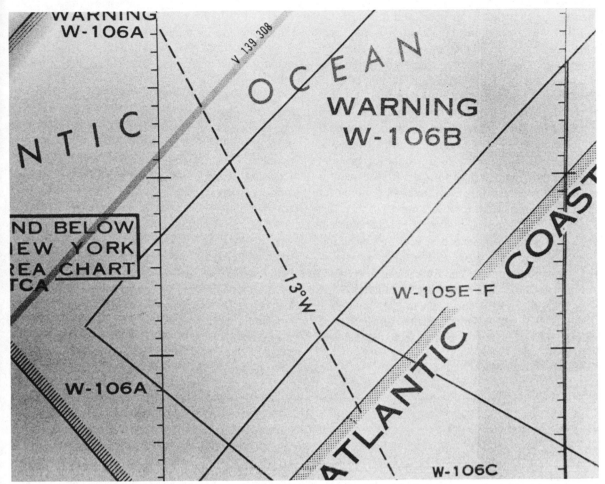

Fig. 1-10. Magnetic Variation is shown on Sectional charts by dashed, magenta colored Isogonic Lines. Magnetic North is 13 degrees west of True North in an area along and to either side of this particular "ISO" Line.

time and fuel. Both time and fuel depend on Ground speed, so Ground speed must be computed first. Ground speed is also computed on the wind side of the DR computer; with the computer still set to a TAS of 147 knots, TC of 274, and winds of 330 degrees True at 30 knots, we find that Ground speed should be 128 knots. (The tailwind component is therefore negative 18 knots: $128 - 146 = [-18]$, which is the same as a headwind of 18 knots.)

Once we have an estimated ground speed, we can go back to the slide rule side of the DR computer and use this figure to determine estimated time between fixes (ETE, or Estimated Time En route). For instance, the distance between the departure point and

Fix #1 is 14 nautical miles. With 128 over the pointer, we rotate the cursor to 14 on the top scale and read time on the inner scale: 7 minutes, after rounding to the nearest minute. (Not all computers have cursors. If yours does not, you will have to align the appropriate figures visually—a pencil helps.) Since we have estimated the ground speed to be constant for the duration of this short trip, the pointer can be left on 128, and the time between segments for the rest of the trip can be estimated simply by rotating the cursor to each of the respective distances and noting the time underneath. Each of these times has been entered in the ETE column and the column itself has been totaled to give total time en route at the top (in minutes, and then beside that figure, in hours and minutes).

The only remaining computation is fuel consumed per segment and total fuel required. These estimates are done in exactly the same way as the time estimates: fuel flow is placed over the pointer (11.5 GPH) and the cursor is set over estimated time between fixes (from the previous computations). Fuel required per segment is then read under the cursor on the outer scale. Total fuel on board is entered at the top right hand corner of the flight log portion, and fuel remaining at the end of the first segment is estimated by subtracting estimated fuel required for the first segment from total fuel onboard, then fuel required for the second segment from that amount, and so on until all segments have been accounted for.

The last estimated fuel remaining figure represents reserve fuel. The FAA (Federal Aviation Administration) requires at least 30 minutes of reserve fuel under VFR during the day, 45 minutes at night (FAR 91.22.). Thirty minutes of fuel can be determined by leaving the pointer set to the fuel flow

rate of 11.5 (in this case), and the amount—5.75—is then read over 30 minutes (or 8.6 for 45 minutes). Since we estimate that we will have 35.1 gallons remaining at our destination, our fuel reserves are well within legal limits, day or night.

The preflight navigational planning for this VFR flight relying entirely on pilotage and dead reckoning is now completed. During the flight itself, the preplanned flight log becomes an invaluable navigational aid. The actual time of departure is noted and recorded under Time Off (Fig. 1-11) and, when workload permits, the estimated times en route (the numbers in the ETE column) are added to the Time Off figure to obtain initial ETA (Estimated Time of Arrival) for each segment. The actual time between fixes is entered for each segment under ATE (for Actual Time En route), and after one or two segments have been completed, an actual ground speed (ACT GS) can be determined using the DR computer: Distance over Actual Time En Route equals Actual GS over the pointer. If the ACT GS differs significantly

```
E-6B COMPUTATIONS
        FUEL            WIND   WIND                      VAR
   TAS  FLOW    TC      DIR    KTS    WCA      TH       -E +W     MH     DEV    CH
-----------------------------------------------------------------------------------
   147  11.5    274     330    30     10       284      14        298    2      300
   KTS  GPH                                    R                  W
===================================================================================
FLIGHT LOG
   DEPT                 TOTAL                  TOTAL                    TIME               FUEL ON
   POINT                DIST                   TIME                     OFF                  BOARD
:-----:------------:------:------------:------:------:HOURS : MINS :------:------------:------:------:
   MVY                   99                     47      0      47    0905 :                    44
:-----:------:------:------:------:------:------:------:------:------:----ESTIMATED---:ACTUAL:
: FIX : HDG : DIST : REM :EST GS :ACT GS: ETE : ATE : ETA : ATA : FUEL : REM : REM :
:-----:------:------:------:------:------:------:------:------:------:------:------:------:
   1    300    14     85    128   : 105 :  7  :  8  : 0912 : 0913 : 1.3  : 42.7 : 42 :
:-----:------:------:------:------:------:------:------:------:------:------:------:------:
   2    300    22     63    128   : 120 : 10  : 11  : 0922 : 0924 : 2.0  : 40.8 : 40 :
:-----:------:------:------:------:------:------:------:------:------:------:------:------:
   3    300    31     32    128   : 125 : 15  : 15  : 0937 : 0939 : 2.8  : 38.0 : 37 :
:-----:------:------:------:------:------:------:------:------:------:------:------:------:
   4    300    19     13    128   : 125 :  9  :  9  : 0946 : 0948 : 1.7  : 36.3 : 35 :
:-----:------:------:------:------:------:------:------:------:------:------:------:------:
  MMK   300    13      0    128   : 157 :  6  :  5  : 0952 : 0953 : 1.2  : 35.1 : 34 :
:-----:------:------:------:------:------:------:------:------:------:------:------:------:
:-----:------:------:------:------:------:------:------:------:------:------:------:------:
:-----:------:------:------:------:------:------:------:------:------:------:------:------:
:-----:------:------:------:------:------:------:------:------:------:------:------:------:
```

TC + WCA + VAR + DEV = CH

Fig. 1-11. This is how the VFR log for the flight between MVY and MMK might look at the completion of that flight. Hand lettered figures represent data entered enroute.

from estimated ground speed, new ETEs can be computed and new ETAs recorded. Actual fuel remaining can also be compared to estimated fuel remaining over each fix. If fuel is being consumed at a faster rate than estimated, then appropriate action to ensure adequate reserves should be taken prior to its becoming a problem (i.e., reduce power to a more economical setting, or make a precautionary fuel stop).

Compass heading is corrected with pilotage: the pilot visually observes the aircraft path over the ground and modifies the compass heading as necessary to maintain the course desired. Since compass headings are computed on the basis of forecast estimates of winds aloft, pilotage is an essential component of VFR dead reckoning. For this reason, VFR navigation based entirely on dead reckoning should only be attempted when the ceiling and visibility are good enough that the planned visual checkpoints can be seen from a distance of several miles.

Figure 1-11 shows how this flight log might look at the completion of the trip. Hand-lettered figures represent the data that would be entered en route as the flight progressed. Note that the ACT GS for the first segment was somewhat slower than planned. This is probably because we did not specifically allow for reduced airspeed during climb-out. For short trips this small inaccuracy is acceptable, especially since the time and fuel lost will usually be regained on the descent. For longer trips, with more extended climb segments, it is wise to compute a separate ground speed estimate for the climb segment based on average TAS during the climb.

Fuel remained well within limits at each point, as can be seen by comparing the Actual REM fuel figures to the Estimated REM fuel calculations. Actual fuel remaining appeared to be slightly less than estimated, but as a practical matter, it would be difficult to read many fuel gauges to anything like the accuracy of the estimated fuel remaining figures, and, in this case, actual fuel remaining was essentially the same as estimated.

Planned carefully and conservatively, dead reckoning can be a very effective form of area navigation for VFR operations. It has fallen somewhat out of favor in the last 30 years, due mainly to the ease of use and reliability of VOR navigation, but, for the recreational pilot in particular, dead reckoning offers an enjoyable, economical, and rewarding method of navigation.

IFR APPLICATIONS

Dead reckoning is seldom used for primary navigational guidance under IFR (although there are exceptions, as we will see); dead reckoning simply isn't accurate enough to provide positive separation and course guidance without visual verification. Actually, it isn't dead reckoning that is at fault, but the inadequacy of the winds aloft data. If winds aloft data could be obtained that were continuously accurate to within a couple of degrees and a knot or two, then dead reckoning might be accurate enough for IFR operations, but that simply is not practical.

Dead reckoning is, however, still used extensively in IFR operations for preflight planning, for en route estimates, and to supplement other, more precise forms of navigation. In fact, dead reckoning is such an integral part of IFR operations that pilots frequently do not even realize that they are using dead reckoning.

Figure 1-12 shows an example of an IFR flight log that was generated on a personal computer with spreadsheet software. (The use of computers is certainly not necessary to IFR flight planning, and there is nothing on this flight log that could not be derived manually using a DR computer, but the computer adds speed and accuracy, and the spreadsheet software makes it very easy to experiment with different flight parameters.) This particular flight between Boston, Massachusetts (BOS) and Charleston, South Carolina (CHS) happens to be based on the Cessna Citation, a well-known corporate jet; the principles illustrated, however, are the same for all IFR aircraft types.

We will not go over this flight log in the same detail we went over the VFR flight log, but the key differences are worth noting. First, for IFR operations on Victor or Jet Airways (still by far the most common kind of IFR navigation), wind correction angles are normally not computed. There is nothing wrong with doing so, but, for the experienced IFR pilot, wind correction angles can be determined very

```
FLIGHT LOG:  BOS   TO   CHS          ALT: FL 350 POWER: MAX CRUISE          FUEL LOAD:   3600
                                     TAS:  346        FUEL FLOW:  905  PPH               POUNDS
------------------------------------------------------------------------------------------------
  DEP                      TRIP       TRIP TIME                TIME                  TRIP
  POINT                    DIST        2   HOURS               OFF                   FUEL
  BOS                      728        37   MINS           :       :                  2367
================================================================================================
             WINDS  EST          DIST                ACTUAL                    FUEL  FUEL  ACTUAL
  LEG   FIX  (-HW)  GS   DIST  REMAIN  ETE    ATE     GS    ETA    ATA         REQ   REM  FUEL REM
------------------------------------------------------------------------------------------------
  #1   BOSOX  -40  210   24    704     7   :_____:_____:_____:_____:      103  3497  :_____:
  #2   PUT    -50  200   22    682     7   :_____:_____:_____:_____:      100  3397  :_____:
  #3   JFK    -70  276  119    563    26   :_____:_____:_____:_____:      390  3007  :_____:
  #4   CYN    -70  276   57    506    12   :_____:_____:_____:_____:      187  2820  :_____:
  #5   ATR    -70  276   71    435    15   :_____:_____:_____:_____:      233  2587  :_____:
  #6   SBY    -70  276   31    404     7   :_____:_____:_____:_____:      102  2485  :_____:
  #7   CCV    -70  276   65    339    14   :_____:_____:_____:_____:      213  2272  :_____:
  #8   ORF    -70  276   29    310     6   :_____:_____:_____:_____:       95  2177  :_____:
  #9   CVI    -60  286   45    265     9   :_____:_____:_____:_____:      142  2035  :_____:
  #10  ISO    -50  296   69    196    14   :_____:_____:_____:_____:      211  1824  :_____:
  #11  CRE    -50  296  110     86    22   :_____:_____:_____:_____:      336  1488  :_____:
  #12  CHS    -40  306   86      0    17   :_____:_____:_____:_____:      254  1233  :_____:
  #13                                      :_____:_____:_____:_____:                 :_____:
  #14                                      :_____:_____:_____:_____:                 :_____:
  #15                                      :_____:_____:_____:_____:                 :_____:
  #16                                      :_____:_____:_____:_____:                 :_____:
  #17                                      :_____:_____:_____:_____:                 :_____:
  #18                                      :_____:_____:_____:_____:                 :_____:
  #19                                      :_____:_____:_____:_____:                 :_____:
  ALTN: SAV  -10  336   70      0    13   :      :      :      :      :      189  1045  :       :
------------------------------------------------------------------------------------------------
  IFR FUEL REQ               :
    DEST        2367         :
    ALTN         189         :
    DELAYS       200         :
    45 MINS      679         :
  ----------- ------         :
    TOTAL       3434         :
                            :
_____:
```

Fig. 1-12. An IFR flight log for a flight in a corporate jet between BOSTON and CHARLESTON, as it might look prior to flight.

rapidly in flight using VOR information, and it therefore isn't necessary to calculate them ahead of time.

Most experienced IFR pilots also do not need a step-by-step flight planning worksheet. They are familiar with the difference between True and Magnetic North, and between TAS and IAS, they know how to estimate a tailwind component from winds aloft data, and they can work with tailwind components and TAS directly to estimate ground speed. For IFR operations along Victor Airways, courses are already described in terms of Magnetic North, so that conversion is eliminated. For these reasons, the typical IFR flight log omits the E-6B section common to VFR flight logs.

The most important difference between flight planning for IFR and for VFR operations is that fuel planning is generally a much more important factor in IFR operations than it is in VFR operations. VFR flights tend to be over short distances with full tanks

of fuel; IFR flights tend to cover longer distances than VFR flights, and reserve fuel capacity therefore tends to be less. In addition, the IFR pilot has less flexibility when fuel is a problem—the VFR pilot may have visual contact with alternate airports over a good part of his route of flight, but the IFR pilot may have to fly to and complete an approach at a suitable IFR airport in order to obtain fuel. Even if the IFR pilot arrives at his destination with adequate fuel reserves, he may find that the weather precludes his landing there and must proceed to and complete an approach at an alternate airport. All of these factors make fuel planning for IFR more critical than for VFR. (Ironically though, the VFR pilot is statistically more likely to run into problems with fuel than the IFR pilot is.)

In VFR flight planning, unless the flight is an unusually long one, winds aloft are normally averaged over the entire route of flight. However, in IFR flight planning, unless the flight is an unusually short one, or one in which the tailwind component is constant for the entire flight, a tailwind component (plus or minus) is normally computed for each leg of an IFR flight. This results in much more accurate ground speed estimates (particularly if reduced TAS during climb-out is also accounted for), which in turn results in more accurate time en route and fuel required estimates.

In this example, in the Winds column we show a tailwind component of negative 40 initially (a headwind), increasing to negative 70 over the JFK VOR, holding steady at negative 70 for several segments, and then gradually dropping off to negative 40 again over Charleston. Estimated Ground speeds, based on a TAS of 250 during the climb and 346 en route, vary from 210 knots to 306 knots. Estimated time between route segments and estimated fuel required for each segment has been computed using the ground speed estimate appropriate to each segment.

Total estimated fuel required, including fuel to fly to an IFR alternate, fuel for 45 minutes at normal cruise (FAR [Federal Aviation Regulation] 91.23); and a fuel allowance for delays, has been computed at the bottom of the flight log (3,434 pounds). Since total fuel on board is 3,600 pounds (top right), the fuel load would seem to be adequate in this case. As the flight progressed, the remainder of the flight log would be filed in: ATE, ACT GS, ETA for each segment, ATA (Actual Time of Arrival) over each fix, and actual fuel remaining (Actual REM) over each fix. (See Fig. 1-13.) This allows actual fuel remaining to be compared to estimated fuel remaining at any point, and an alternate plan of action initiated if actual fuel remaining appears to be consistently dropping below estimated fuel remaining. Actual fuel flow can also be calculated (although this flight log does not have a specific block for it) with the DR computer: actual fuel consumed on the outer ring goes

```
FLIGHT LOG:  BOS   TO   CHS        ALT: FL 350 POWER: MAX CRUISE        FUEL LOAD: 3600
                                   TAS:  346      FUEL FLOW:  905  PPH           POUNDS
--------------------------------------------------------------------------------------
DEP                    TRIP       TRIP TIME          TIME          TRIP
POINT                  DIST        2   HOURS         OFF           FUEL
BOS                    728        37   MINS         :1400:         2367
======================================================================================
            WINDS  EST        DIST             ACTUAL              FUEL  FUEL    ACTUAL
LEG  FIX    (-HW)  GS   DIST REMAIN ETE  ATE    GS   ETA   ATA     REQ   REM   FUEL REM
--------------------------------------------------------------------------------------
#1   BOSOX  -40    210   24   704    7  :  7 : 210 :1407:1407:    103   3497 :  3500  :
#2   PUT    -50    200   22   682    7  :  7 : 200 :1414:1414:    100   3397 :  3400  :
#3   JFK    -70    276  119   563   26  : 24 : 298 :1440:1438:    390   3007 :  3050  :
#4   CYN    -70    276   57   506   12  : 11 : 310 :1452:1449:    187   2820 :  2900  :
#5   ATR    -70    276   71   435   15  : 15 : 284 :1507:1504:    233   2587 :  2650  :
#6   SBY    -70    276   31   404    7  :  7 : 264 :1514:1511:    102   2485 :  2550  :
#7   CCV    -70    276   65   339   14  : 15 : 260 :1529:1526:    213   2272 :  2300  :
#8   ORF    -70    276   29   310    6  :  7 : 260 :1534:1533:     95   2177 :  2200  :
#9   CVI    -60    286   45   265    9  :  9 : 285 :1543:1542:    142   2035 :  2050  :
#10  ISO    -50    296   69   196   14  : 14 : 296 :1557:1556:    211   1824 :  1850  :
#11  CRE    -50    296  110    86   22  : 21 : 315 :1619:1617:    336   1488 :  1500  :
#12  CHS    -40    306   86     0   17  : 20 : 258 :1636:1637:    254   1233 :  1200  :
#13                               :     :     :    :    :        :     :         :
#14                               :     :     :    :    :        :     :         :
#15                               :     :     :    :    :        :     :         :
#16                               :     :     :    :    :        :     :         :
#17                               :     :     :    :    :        :     :         :
#18                               :     :     :    :    :        :     :         :
#19                               :     :     :    :    :        :     :         :
ALTN: SAV   -10    336   70     0   13  :     :     :    :    :    189   1045 :         :
--------------------------------------------------------------------------------------
IFR FUEL REQ                  :
   DEST    2367               :
   ALTN     189               :
   DELAYS   200               :
   45 MINS  679               :
--------- ------              :
   TOTAL  3434                :
                             :
```

Fig. 1-13. This is how the flight log shown in Fig. 1-12 might look at the completion of the flight.

over time on the inner ring, rate can then be read over the pointer. This is a good double-check on the power setting and the fuel flow indicator.

All of these computations—estimated ground speed, time en route, time of arrival, fuel en route, fuel remaining, and actual ground speed and fuel flow—are important to safe IFR operations, and all are based on fundamental principles of dead reckoning. Regardless of the type of navigational system being used for primary course guidance, dead reckoning is an essential element of en route navigation.

DEAD RECKONING IN REMOTE AREAS AND OVER WATER

There are times when dead reckoning must be used for primary IFR navigational guidance, although these times are rare. When operating in remote areas (or, sometimes, over water), where ground-based navigational aids are limited or unreliable and the aircraft is not equipped (and not required to be equipped) with a long-range navigational system, navigation by dead reckoning is necessary. These kinds of situations are hard to find, but they do still exist and the pilot who anticipates operating in remote areas or over water should be prepared for them. (Pilots who routinely operate in remote areas of the world, or over water, usually operate aircraft equipped with long-range navigational systems; it is the non-routine operation in remote or over water areas we are most concerned with here.)

In those cases where dead reckoning is required, it will seldom be required over the entire route of flight—it is much more likely to be confined to an intermediate segment outside of the range of navaids at the departure and destination ends of the route. This enables an accurate ground speed and wind correction angle to be established initially, allowing the dead reckoning portion to be accomplished with considerably greater accuracy than it would be if the pilot had to rely on estimated winds aloft data alone. This also provides for a certain margin of error, since accurate navaids can presumably be picked up nearing the destination and "homed in on". In these circumstances, dead reckoning provides the bridge between navaids.

EMERGENCY INSTRUMENT NAVIGATION

Dead reckoning is the ultimate back-up form of navigation. In the event of total electrical failure (the failure of all generators or alternators and the depletion of all battery sources), or in the event of a fire in the radio rack or other emergency negating the use of all communication radios and navigational aids, the only form of navigation left is dead reckoning. Even with nothing but a compass, a clock, and an airspeed indicator (actually, even the airspeed indicator can be dispensed with if power is held constant and TAS was known prior to the emergency), emergency navigation to visual conditions can often be accomplished with a fairly high degree of certainty, the exact degree depending upon the accuracy of information available at the time dead reckoning was begun, and the consistency of the winds aloft.

The pilot who experiences a total electrical failure (or total communication and navigational failure) while in Instrument Meteorological Conditions (IMC) will want to reach visual conditions as rapidly as possible, both to obtain more positive navigational guidance, and to minimize the risk of collision. If he has been maintaining an accurate and current flight log, he should have a good idea of his position at the time of the emergency, as well as an accurate ground speed, wind correction angle, and rate of fuel flow. From his preflight weather briefing he should also be aware of the nearest area with visual conditions (scattered to broken clouds for all altitudes), or at least be aware of an area with ceilings high enough to risk a descent through the bases.

With this information he can plot a magnetic course from his last known position to the area selected, ensuring that there is no high terrain intervening. A wind correction angle can be determined by modifying the previous WCA in an amount appropriate for the course change, and a new ground speed estimated to reflect the change in heading.

Once the pilot has settled on a heading, the time should be noted and, using the best available ground speed information, estimated time to reach visual conditions calculated. The pilot should also make an estimate of time to fuel exhaustion. The ETA for arrival in visual conditions helps prevent unnecessary, emergency let-downs inspired by panic, and the estimate of time to fuel exhaustion gives the pilot an idea of the amount of time remaining to search for visual conditions. If visual conditions cannot be found and fuel is nearly depleted, then the pilot will have to fly to the area in range with the highest ceilings and the flattest terrain (large bodies of water are often good choices), and make a slow, controlled let-down until ground contact is made and a suitable emergency landing site found.

In these difficult circumstances, dead reckoning not only offers the best chance to find visual conditions, but it is also the best (indeed, the only) way to keep track of estimated present position—what navigators call ''Most Probable Position'' (MPP). Keeping track of MPP is the best defense against flying in circles, and it also offers the best chance of survival if a descent through the clouds becomes necessary. (An important procedural matter once on the ground is to immediately call the nearest Flight Service Station and inform them of your position and condition, and cancel your IFR flight plan—ATC will be nearly as relieved to find out that you are on the ground as you no doubt will be.)

ADVANCED APPLICATIONS

Because dead reckoning is nothing more than the practical application of fundamental Laws of Motion to navigation, dead reckoning is basic to all navigation, regardless of how simple or complex the equipment or method. Dead reckoning capability is, in fact, built in to many LORAN-C and OMEGA units (long-range nav systems covered in Chapters 9 and 10); when primary navigational information is lost, the long range unit automatically reverts to a dead reckoning mode based on stored information: Last Known Position, Ground speed, Winds Aloft, Course. In fact, the most complex and sophisticated navigational system of all, the Inertial Navigation system (covered in detail in Chapter 11), is actually nothing more than a very fast dead reckoning computer coupled to extremely sensitive motion detectors, capable of accuracies on the order of one nautical mile per hour of operation, and able to function completely independently from any ground facility.

CONCLUSION

There is nothing inherently inaccurate about dead reckoning; dead reckoning is limited only by the information provided. When part of the information provided is estimated winds aloft based on extrapolations from a small number of actual observations, then dead reckoning is not terribly accurate (but it may be accurate enough for certain applications). When the information provided is based on actual, observed conditions in flight, such as from VOR/DME data, or long range navigational system data, then dead reckoning can be quite accurate, at least for short periods of time. When the information is highly sensitive rate information about all three axis, able to detect any deviation from an initial position (inertial systems), then dead reckoning is extremely accurate, limited only by the amount of time since the system was initiated, and the accuracy of the sensors and timing devices installed.

It is entirely appropriate that the first navigational system a pilot learns is dead reckoning. Dead reckoning is a part of every form of navigation, from recreational VFR flying to long range, over-water navigation. Dead reckoning is also the one system that will always work, requiring only a functioning compass, clock, airspeed indicator, and a pilot who understands its principles. For that reason alone, it may also be the most important system a pilot learns.

Chapter 2

VOR Navigation Fundamentals

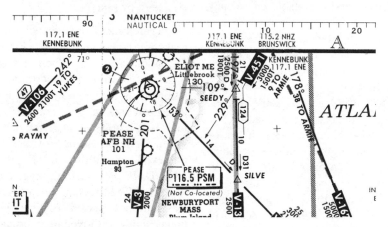

The VOR system of land-based navigation for aircraft has been the workhorse of the U.S. navigational and instrument system since the 1950s. VORs can now be found in virtually every part of the world, proving the virtues of this simple but highly effective navigational system from the pre-transistor era. If the DC-3 is the aircraft that proved to the general public flying could be safe, comfortable and reliable, then the VOR is the navigational system that made transportation by air not only safe, but routine.

Prior to the implementation of the VOR system, instrument navigation—navigation without any outside reference to terrain—was the exception rather than the rule, even for airline and military operations. Homing beacons were used for instrument navigation, along with aural, four-course ranges. Both were slow, difficult to use, and relatively inaccurate.

The VOR system not only made routine instrument navigation possible for all pilots, but it also freed the pilot operating strictly under Visual Flight Rules from the uncertainties of pilotage and dead reckoning. The VOR system provided the general aviation pilot with the ability to navigate easily, and with great accuracy and predictability—important factors in making general aviation practical and popular.

VOR navigation is not without its limitations, and it can be expected to be slowly replaced by newer technologies in the years ahead, but VOR navigation can be expected to continue as the mainstay of air navigation for at least several more years. The US National Airspace System (NAS) is officially based on the VOR system, and all aircraft, including the newest and most sophisticated airliners and corporate jets continue to be equipped with VOR receivers for their primary, land-based navigational systems. The VOR system is the primary navigational system worldwide, and one that deserves thorough study by all pilots.

PRINCIPLES OF OPERATION

The basic principle of operation behind the VOR is very simple: the VOR facility transmits two signals at the same time. One signal is constant in all directions, while the other is rotated about a point. The airborne equipment receives both signals, looks (electronically) at the difference between the two signals, and interprets the result as a *radial* from the station. If the two signals are "in phase", (i.e., match each other), then the VOR receiver says that the aircraft is on the 360 degree radial. If the signals are out of phase by half, then the receiver says that the aircraft is on the 180 degree radial, out of phase by three quarters, the 270 degree radial, and so on. Thus any individual aircraft at any point around the VOR transmitter will receive a unique combination of signals—a combination that corresponds to its position in relation to the VOR station.

Probably the simplest way to visualize the way VOR works is to think of the VOR as transmitting 360 different signals radiating out from the station. The aircraft, however, only receives one of those signals—the one that corresponds to its position from the VOR. (See Fig. 2-1.)

It is also important to remember that the VOR system provides only one piece of information (but an important piece): it tells the pilot along which radial his aircraft is located. It does not tell him his position on that radial—he could be right next to the station, many miles away, or something in between—nor does it tell him anything about the direction he is headed. (See Fig. 2-2.)

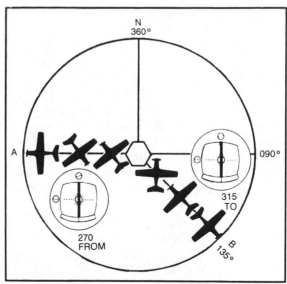

Fig. 2-2. The aircraft located on line A are all on the 270 degree radial; the aircraft located on line B are all on the 135 degree radial.

It is important to remember that the information which the VOR supplies is completely independent of aircraft heading. The VOR simply says "You are presently located somewhere along the following radial, and that is all I can tell you." This may sound contradictory to those with some experience in actually using the VOR system, but it is nonetheless true and is an important point to keep in mind, one that will become clearer as we get into VOR applications in more detail. (The manner in which that information is interpreted may depend upon aircraft heading, but the VOR itself transmits only one bit of information, and that is the present radial from the station.)

IDENTIFICATION

Tuning the wrong station will obviously result in serious errors, and simply double-checking the frequency selection is not sufficient safeguard in itself against such errors. More importantly, even if the proper frequency has been tuned, this still does not ensure that a valid signal is being received unless the station is properly identified.

There are several reasons for this. The FAA tries to space VOR stations around the country in such a way that stations having the same frequency are far

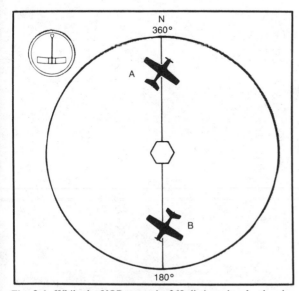

Fig. 2-1. While the VOR transmits 360 distinct signals, the aircraft receives only the signal that corresponds to its radial from the VOR station: Aircraft A is located on the 360 degree radial; Aircraft B is located on the 180 degree radial.

enough apart that they won't interfere with each other, but this isn't always possible. As aircraft altitude increases, more stations can be received, and sometimes, because of peculiar terrain characteristics, the station farthest away will be received, rather than the one desired or expected. A check of the station identification will reveal that the station being received is not the correct one. In other cases, particularly when navigating directly to a fairly distant VOR, there may be more than one station with the same frequency within receiving range. Without proper identification, the pilot could fly to the wrong station.

In addition to interference errors, certain VOR maintenance procedures require that a VOR signal be transmitted. While this work is being done, there is no assurance that the signal is correct. In these sit-uations, the maintenance personnel turn the identifying signal *off* to indicate that a valid signal is not being transmitted. The signal transmitted may be close enough to seem correct—in fact, it may actually be correct—but until it has been tested, it still cannot be relied upon. Thus the absence of an identifying signal indicates that the VOR in question cannot be relied upon.

The actual process of tuning the station and checking the identification is simple. Each station has a four or five digit frequency and a three letter identifier; the identification signal itself is a transmission of those three letters in Morse Code. (It isn't necessary to know Morse Code; the code ident symbols are printed on the navigational charts, along with the frequency and station identification letters. See Fig. 2-3.) After entering the proper frequency into the

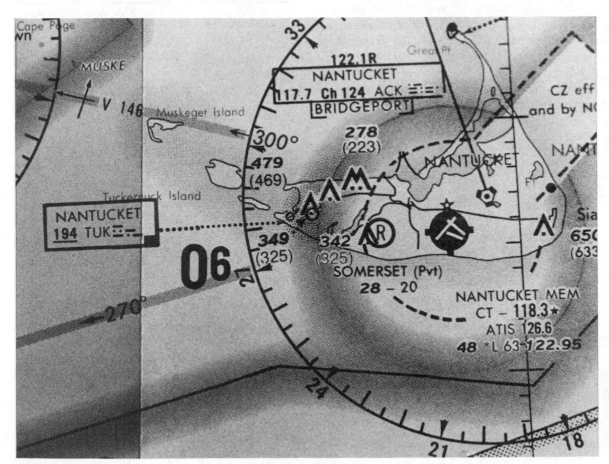

Fig. 2-3. All of the information necessary to tune and identify the VOR station is provided on the charts.

Fig. 2-4. Two VOR receivers, with their associated Course Deviation Indicators (CDIs) above. Photo courtesy of Narco Avionics.

VOR receiver (Fig. 2-4), the volume control should be turned up in order to hear the Morse Code identification. Most receivers also have an ident feature ("Pull Ident" in Fig. 2-4), which is simply a filter that allows the ident to be heard better.

Some VOR facilities also transmit a voice identification. By whatever method, until positive identification is made, the VOR information cannot be trusted.

ORIENTATION

There are two ways in which VOR information can be displayed in the cockpit. One is the Radio Magnetic Indicator (RMI), and the other is the Course

Deviation Indicator (CDI). Each displays radial information from the VOR, but in different ways. There are advantages and disadvantages to each, and the ideal cockpit arrangement is to have both. Since the CDI is the more common display (in fact most pilots call the CDI simply, "the VOR"), that is the type of display referred to in this chapter. (We will save our discussion of RMI indicators for the chapter on VOR/DME applications. At that point, VOR systems will have been covered thoroughly enough that it will only be necessary to examine the differences between CDIs and RMIs.)

The main advantage to the CDI over the RMI is that it is easy to use. The main disadvantage is that it requires orientation to be used correctly. This is quite easy in practice, but can be quite confusing in theory (as the people who design FAA written tests seem to have discovered, no doubt to their delight).

Figure 2-5 illustrates a typical VOR CDI. In order to determine along which radial the aircraft is positioned, the Omni Bearing Selector (the knob marked "OBS") is rotated until the triangular arrow points to FR (for From) and the vertical needle has centered. (Some CDIs use windows with To and From signs instead of arrows pointing to symbols.) The pointer at the top of the dial then indicates the correct radial. In this case the CDI in Fig. 2-5 is indicating that the aircraft lies somewhere along the 254-degree radial. By referring to the appropriate aeronautical chart, and then extending a line from the station outwards along the radial indicated using the compass rose conveniently provided on the chart, the aircraft position can be located somewhere along that line. (See Fig. 2-6.) This is called a *Line of Position* and is commonly abbreviated (and called) an LOP.

The concept of a Line of Position or LOP—

Fig. 2-5. When the needle is centered with a From indication, the aircraft position along a specific radial is indicated by the pointer at the top—the 254 degree radial in this case. Photo courtesy of Collins Avionics, Rockwell International.

knowing where you are somewhere along a line, but not the exact point along that line—is one which we will refer to again and again in this book. It is one of the most fundamental principles of navigation.

Turning the OBS until a From indication appears and the needle has centered will indicate aircraft position along a given radial, regardless of which way the aircraft is headed at the time. The aircraft could be pointed directly at the VOR station, directly away from it, or along side it, and the result would be the same. At the moment the needle centers, the aircraft is on the radial indicated. (Naturally, if the aircraft is flying by the station, in the next moment or two it will be on another radial, and the OBS will have to be used to re-center the CDI to indicate the new radial.)

By itself, this is all a VOR station can tell you— along which radial you are currently positioned. This is useful information: half of what you need to know to find out exactly where you are for instance (the other being the distance from the station along that radial.) But the beauty of the CDI is that it allows you to interpret that information so that you can fly to the VOR station, over the station, and then from it. The CDI can do more than just tell you along which radial you are located: it can also be used to actually go somewhere.

To go directly to the VOR station from any given position, orient the CDI by turning the OBS until the arrow points to the To symbol and the needle has again centered. The number at the top now indicates the magnetic bearing that will take you to the station. All you have to do, initially, is turn the aircraft to the magnetic bearing indicated and fly there.

When you arrive over (or abeam) the station, the arrow will disappear and an Off sign will appear, and then, as you fly away from the station on the other side, the arrow will again appear and point to From. What the To/From pointer does then, is divide the area around the VOR into two halves: a To side, a From side, and an Off line that divides the two. (See Fig. 2-7.) Once the CDI is oriented either To or From the station, as long as the aircraft remains on the To side, a To will show in the window. When it crosses the boundary separating the two sides it will show Off, and once on the far side, it will show From.

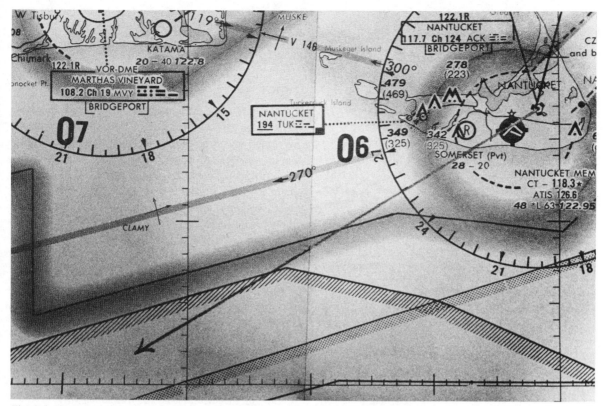

Fig. 2-6. If a line is drawn outward from the VOR along the radial indicated by the VOR Course Deviation Indicator, the aircraft position will fall somewhere along that line. This is called a Line of Position.

The CDI information will be meaningful and accurate only to the extent the CDI is oriented properly—the CDI has no brain and it has no idea which direction the aircraft is headed. If the aircraft is headed away from the VOR and the needle has been centered but with a To indication in the window, all it is saying is that that bearing is the proper radial to fly To the station, whether you want to or not, and regardless of which way it is headed at that moment.

We have all heard the expression "All roads lead to Rome", and likewise, all radials lead To the station. Therefore, to go To the station all you have to do is center the needle with a To indication and fly in that direction. But when the time comes to leave Rome, "all" roads won't do; you have to take the correct road if you want to go to a specific destination. Navigating from a VOR is exactly the same situation: there are 360 potential roads, or radials, outbound from any VOR, and only one will point you

in the direction you want to go. So while going To a VOR from any given position is a simple matter of centering the needle with a To indication, once over the VOR you must select the desired course outbound.

Depending on the situation, one of two methods can be used to determine the radial desired. The first method is to use the published radials of the *Victor Airway System*. The Victor Airways are a predetermined set of radials linking VOR stations in a network of routes. The radials that comprise this network are printed on both the VFR and IFR Charts. (See Fig. 2-8.) Once over the VOR you simply look on the chart for the outbound route, and the correct radial will be printed on the chart.

The second method involves drawing a line on the chart from the VOR in question to the next fix. (See Fig. 2-9.) The next fix could be anything you wanted, but logically would be either another VOR

25

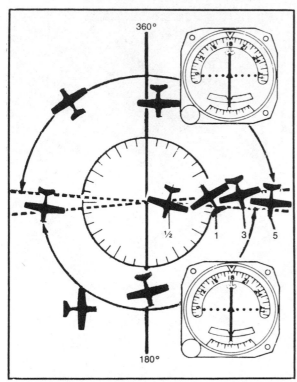

Fig. 2-7. With the CDI set to 180, the To/From pointer divides the area around the VOR into two halves: a To side to the north of the VOR, and a From side to the south of the VOR. An Off area separates the two.

or the destination airport. The correct radial can then be read either directly off the chart using the compass rose around the VOR, or if more accuracy is desired, the radial can be measured with a plotter.

Assuming the CDI was oriented correctly with a To indication inbound to the VOR, as the aircraft passes over the station to the other side of the VOR the arrow will automatically flip over to a From indication—it isn't necessary in this case to rotate the OBS until the From flag appears. But it is necessary to set the desired outbound radial at the top of the indicator, and then turn the aircraft in that direction.

So to quickly summarize the basic principles of VOR orientation and operation up to this point: to find out at any time along which radial the aircraft lies (a LOP), turn the OBS until the needle centers with a From indication and read the radial under the pointer. To fly To or From a VOR station, first orient the CDI to either a To or a From indication,

respectively. Then, to fly To the station, center the needle and turn in that direction; to fly From the station select the desired radial with the OBS and turn in that direction.

DRIFT

If there were no wind, this is all that would ever be necessary to navigate using VORs. (In fact, if there were no wind, you wouldn't need VORs at all—you could just measure a course and fly in that direction for a predetermined amount of time.) But of course, there always is wind, or at least the possibility of wind, so some means of accounting and correcting for wind is necessary.

Assume that an airplane is headed due north on the 360-degree radial, with a wind blowing directly from the east. As it heads north it will drift from the 360-degree radial first to the 359-degree radial, then the 358-degree radial, then the 357-degree radial and so on. As it does so, the CDI needle will drift off center to the right.

The pilot could find out which radial he was drifting over to as he flew along by re-centering the needle. First it would re-center on 359, then 358, and so on. But this would only tell him along which radial the aircraft was positioned—it wouldn't tell him what to do to correct the situation.

WIND CORRECTION

You can see from Fig. 2-10 that the solution in this case is to turn the aircraft back to the east somewhat. As the aircraft heads back toward its desired northerly course, it will again cross back over the 358- and 359-degree radials, the needle will swing back to the left, and when the aircraft is back on the 360-degree radial the needle will again be centered. This time, however, the needle will be centered because the aircraft was flown back on course, and not because the OBS was turned to center it. By adjusting the heading of the airplane to keep the needle centered, the proper wind correction angle will be automatically determined.

As long as you have a picture, it is fairly obvious which way to turn to get back on course, but without a picture it is not so easy. The CDI also tells you which direction to turn, provided it is oriented prop-

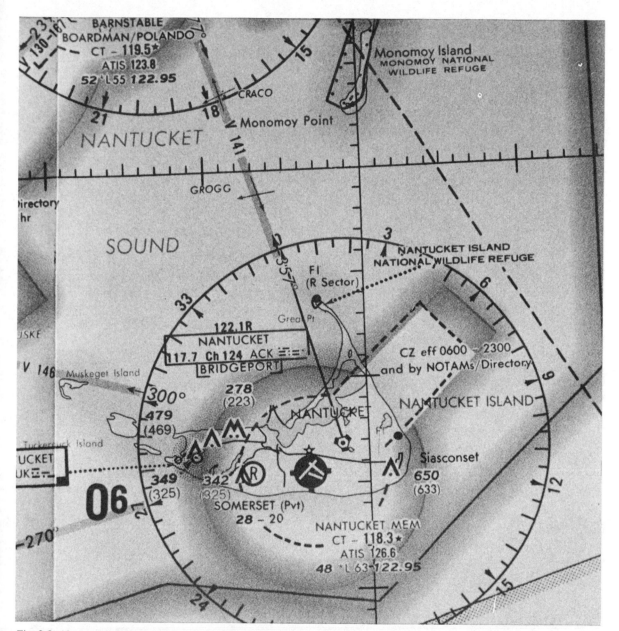

Fig. 2-8. Airway radials are printed on both VFR and IFR charts. VICTOR 146, for instance, on the left hand side of this picture, is based on the 300 degree radial off the NANTUCKET VOR in this segment.

erly; that is, it must be oriented with a From indication when you are indeed flying away from the station, and with a To indication when you are indeed headed toward the station, or it won't work properly.

In the example above, where the aircraft was fly-

ing outbound from the VOR on the 360-degree radial with an east wind, as the aircraft drifted to the 359-degree radial, the CDI needle would have moved to the right of center. The proper direction to correct is also to the right (the east), which means a turn towards the needle—the needle is to the right of cen-

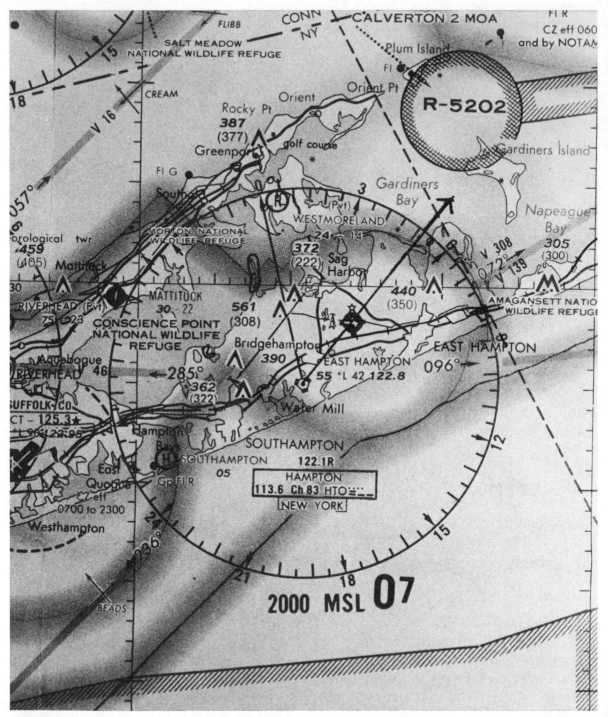

Fig. 2-9. By drawing a line from the HAMPTON VOR to the airport, the correct radial for the airport—52 degrees—can be determined using the compass rose surrounding the VOR.

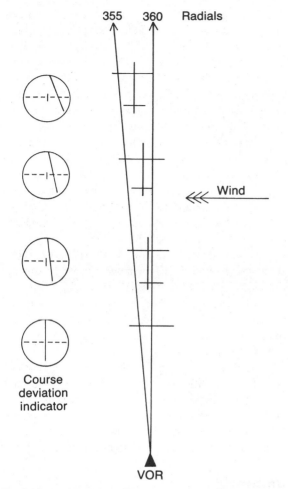

355 360 Radials

Wind

Course deviation indicator

VOR

Fig. 2-10. An aircraft headed due north on the 360 degree radial, with a wind from the East, will drift first over to the 359 degree radial, then the 358 degree radial, and so on. As it does so, the CDI needle will swing to the right. To return to the 360 degree radial, the pilot turns toward the needle.

ter, so you turn slightly to the right. As you correct back towards the 360-degree radial, the needle will return to center.

Thus you do not need to know the direction of the wind, nor do you need to know how hard it is blowing. All you have to do is turn towards the needle; the movement of the needle—further away, back toward the center, or back too far, off center to the other side—will tell you whether the correction has been sufficient or not, and what to do to improve upon

it if it is not. You do exactly the same thing to track To the station, except that the CDI must be oriented so that a To flag shows in the window.

What happens if you orient the CDI incorrectly? That is, what happens if you are flying toward the station, but with a From flag showing in the window? Since the CDI has been told that you are going From the station, when in fact you are headed To the station, it is, in effect, turned around, and will give you all your directions backwards. It will tell you to correct right, for instance, when it "means" left. (It isn't wrong—if you really were going From the station, that would be the proper correction.) Therefore, in this case, if you correct by turning towards the needle, you will be going exactly the wrong way and the needle will drift further away from center—a frustrating situation. This is why proper orientation of the CDI to the direction of flight is so important when using the CDI for VOR tracking.

The designers of the VOR system could have designed the system to work without To/From flags, but then you, the pilot, would have had to remember to turn towards the needle outbound but away from the needle inbound (or vice versa depending on how they decided to design the system.) That would have been very confusing and much more prone to error. I think they did the right thing in including a To/From orientation, but at a price: the CDI must be oriented correctly to work properly.

TRACKING

VOR theory tells you to turn towards the needle to correct for wind drift, but it doesn't tell you how much to turn. Simply turning towards the needle until it starts to re-center and then turning back towards the original heading usually leads to overcorrection and erratic flying. On the other hand, not making a large enough correction in the first place results in the needle drifting even further off center, requiring a larger correction later on. The situation is complicated by three factors. One, the greater the distance from the station, the slower the needle is to react; two, the stronger the crosswind, the greater must be the correction; and, three, the faster the aircraft, the less correction is necessary. "Turn towards the needle'' is fine as theory, but it isn't specific enough in actual practice.

The reason for the first complication is shown in Fig. 2-11. Radials are like thin slices of pie. As the distance increases from the station, the distance between radials increases. Thus a one-degree close in to the station represents a difference of only a few hundred feet between radials, while a one-degree change many miles from the station is equivalent to several miles' separation. Therefore, the greater the distance from the station, the longer it takes to show a shift from one radial to another, and the slower the needle will move.

The reason for the second complication is fairly obvious. The harder the wind is blowing, the more the aircraft will drift and the greater will be the need to compensate. The speed with which the needle moves off center, (taking into account the distance from the station), is a rough indication of the strength of the crosswind, but only a rough indication. It is not precise enough to develop a rule.

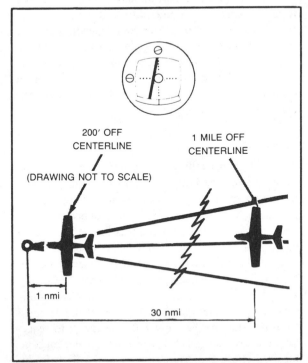

200' OFF CENTERLINE

1 MILE OFF CENTERLINE

(DRAWING NOT TO SCALE)

1 nmi

30 nmi

Fig. 2-11. A one dot deflection at one nautical mile represents a distance off course of 200 feet; the same deflection at 30 nautical miles represents a distance off course of one mile.

The reason for the third complication is less obvious, but not difficult to understand. For any given crosswind and heading, a slower airplane will be blown farther off course than a faster airplane will; the slower airplane will be more affected by the wind. This means that the slower the True Airspeed, the greater the wind correction angle required for any given wind. For instance, an airplane that cruises at 100 knots TAS with a 50-knot direct crosswind will need a 30-degree correction angle to stay on course, while an airplane that cruises at 200 knots will only require a 14-degree correction angle for the same wind. To the extent airspeed varies from climb to cruise to descent, the necessary wind correction angle for any given wind will also vary.

As a result of these complicating factors, there is no way to say exactly how much is the right amount to turn for any given needle deflection. The right amount—the amount that results in a wind correction angle that keeps the aircraft exactly on course, or on track—must be determined in flight by a process called tracking.

There are two basic tracking techniques, both of which are variations on the same theme. The first method, bracketing, works best when the wind direction and velocity is unknown; the second method, estimating, works best when some information as to wind strength and direction is available.

Bracketing

Bracketing, in general, is nothing more than a logical process of trial-and-error in which repeated trials result in increasingly smaller errors. As a result of these repeated trials, a value is eventually found which is the correct one. What separates bracketing from mere guess work is that a logical process is followed so that the number of trials is reduced to the theoretical minimum necessary to determine the correct value.

The first step in any bracketing process is to determine the maximum range expected, try each extreme of that range to see which is closer, and use that result to establish a new range. With each observation the range is narrowed by half, until finally the range is so small that, for all practical purposes, it amounts to a single value.

To bracket a VOR wind correction angle inbound, you start by centering the needle and turning the airplane to match the radial. If there is no wind (or a direct head or tail wind—no crosswind) the needle will stay centered. This is rarely the case.

Assuming there is a crosswind component, after a short period of time you will see the needle begin to drift off center. To bracket a heading (as opposed to just turning towards the needle in some random fashion), you turn towards the needle a predetermined number of degrees. This is your initial, large correction. For most general aviation airplanes, most of the time, a 30-degree heading change will be about the right amount initially.

Having turned 30 degrees towards the needle, you should see the needle start to swing back towards the center as you correct back to the original radial to the station. If the needle doesn't start back towards the center position after your initial correction, you will have to turn further towards the needle, but this will be very rare, indicating an exceptionally strong crosswind.

When the needle reaches the center point again, you cut the initial correction in half—15 degrees in this case. (You can't just sit there with 30 degrees of correction, or the aircraft, and the needle, will continue on over to the other side.) Again you watch the needle. Fifteen degrees might be just the right amount of wind correction angle to keep the needle centered, in which case you have nothing further to do. But the chances are it won't be quite right.

The chances are it will again start to drift off center, either to the right or the left, depending on whether the 15-degree correction was too much or too little. This time you add or subtract (depending on which direction the needle is drifting) half again—seven or eight degrees—to your previous heading. Again the needle should come back to the center, and when it does you cut that correction in half—three or four degrees. You watch for any further drift and add or subtract one or two degrees to the last heading. After having taken half of that out when the needle again centers, you will be within one degree of the exact wind correction angle, which is as close as you can read the compass or directional gyro. If the wind doesn't change, this last heading will take you

directly to the station.

Tracking outbound using the bracketing method is essentially the same procedure, except that the needle cannot be centered using the OBS but must be centered by intercepting (flying over to), the desired outbound radial. After having set the desired outbound radial with the OBS and having turned the airplane in the right direction, wait for the Off flag to be replaced by a From indication and note which way the needle is deflected. Then apply the maximum correction in the direction of needle deflection in order to quickly get back on the desired outbound radial. Once it centers, go back to the heading that matches the outbound radial. At that point you are ready to begin bracketing again.

I said earlier that "for most general aviation airplanes, most of the time", 30 degrees was a good starting correction. The reason for that suggestion—and it is just that, not an absolute—is that most general aviation airplanes with TASs between 100 and 200 knots cruise at altitudes where the crosswind component is generally 50 knots or less. A little experimentation with an E-6B will show that even with a 50 knot direct crosswind, a 30-degree wind correction is enough. In fact, for most winds and airspeeds, the final wind correction angle will be less than 20 degrees. Therefore 30 degrees will almost always be a big enough cut to ensure an adequate bracket. Any correction beyond 30 degrees is probably needle chasing.

Those aircraft that operate in the higher altitudes where the winds are frequently stronger than 50 knots also generally have much higher cruising airspeeds: 200 to 500 knots. A little experimentation will again show that 30 degrees is about the greatest correction angle normally encountered in these situations (and the exceptions will usually be fairly obvious—it is hard to not be aware of a 100 knot crosswind component, regardless of the TAS).

Estimating Wind Correction Angles

Bracketing is the best method to use when the winds aloft are unknown. Regardless of the strength of the winds, or the direction, bracketing will eventually result in the proper wind correction angle, although with a 30-degree initial bracket it may take as many as seven tries before a heading to within one degree of accuracy is nailed down.

When the winds aloft are known with some reasonable degree of accuracy, as they normally would be if the flight has been planned properly or previous legs have been flown, it isn't necessary to bracket every course change, starting from zero each time. Instead, the tracking process can begin from an initial estimate of the wind correction angle, based on either a winds aloft forecast or the wind correction angle from the previous leg.

If, for instance, you have determined from your preflight planning that you can expect a 12-degree wind correction angle to the right on the first leg of your flight, you can save a good deal of experimentation by trying the 12-degree correction initially. If the winds aloft forecast is accurate and your computations are correct, then no further adjustments to the wind correction angle, at least for that leg, will be necessary.

The chances are, however, that the estimate will not be exactly accurate. In this case, the needle will gradually drift off center again. You certainly would not want to take a 30-degree correction at this point, since you have to assume that the initial estimate is at least better than not knowing anything at all about the wind strength, but you don't want to take such a small cut that the needle continues to drift off. Usually something like a five-degree correction will be sufficient to start; if that is not enough, the needle will continue to drift in the same direction, although at a slower rate. In this case you will have to try another five degrees.

In most cases the five-degree correction will be sufficient and the needle will go back to the center; you can then try plus or minus two or three degrees the next time and so on. This process of estimating a wind correction angle, and then correcting the correction, is essentially the same process as bracketing, but it is quicker and more efficient because you start with an initial estimate.

Once you have a wind correction angle nailed down for the first leg, you can then apply that correction angle to the succeeding legs. Unless your course or the winds change substantially, this correction angle will only require small modifications after the initial bracketing.

To summarize tracking techniques: if you don't know the winds, bracket a heading, beginning with a heading that matches the course for one side of the bracket, and plus or minus a given amount (no more than 30 degrees is suggested) for the other. If you have some idea of the wind direction and velocity, try an estimate, and then make small corrections at the first sign of needle drift, gradually bracketing an exact heading. When estimating, the earlier the corrections are made, the smaller they need be and the more accurate and efficient will be your tracking.

VOR tracking becomes second nature with experience. But this will come about only if a conscious effort is made in the beginning to bracket courses to within one degree of accuracy. It takes a little bit of work initially, but it is much better than the frustration of forever chasing needles.

POSITIONING

I said earlier that the VOR station only provides the radial from the station along which the aircraft is positioned. You have already seen how this information can be used to track to or from a VOR station. By combining the information from two stations, you can also use the VOR system to locate your exact position.

The theory behind this is simple. If you know that you are located somewhere along one line, and also somewhere along another line, then your exact location has to be at the intersection of those two lines (since that one point is the only point in common). Thus, if you have readings from two different VORs, each telling you that you are somewhere along a given radial of each, your exact location is defined by the point where those radials (LOPs) cross. (The general rule is that the intersection of two LOPs defines a fix.)

In practice, using two VORs to determine present position is simple. First one VOR is tuned in, the OBS is rotated until a From indication appears (you want to know your position From the VOR), the needle is centered, and the radial noted. Then a second, nearby VOR is tuned in and the procedure repeated. Lines of position can then be drawn outward from each VOR along the radials indicated, and the intersection of the two LOPs indicates the aircraft

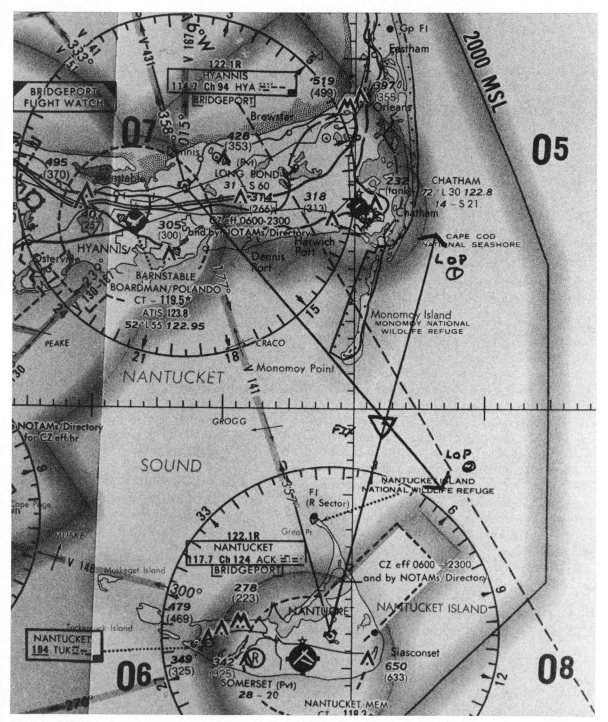

Fig. 2-12. The point where radials, or LOPs, from two different VORs intersect is the aircraft position, indicated in this illustration by the triangular symbol for a fix.

position—a fix. That is the aircraft position at the time the readings were taken. (See Fig. 2-12. Fixes are traditionally marked on navigational charts in the form of a triangle. When three LOPs are used, a triangle is automatically formed by the intersection of those three lines—the smaller the triangle, the better the fix.)

Because this process takes a few seconds, and because the aircraft is presumably moving at the time, the aircraft position will have changed slightly from the time the readings were taken, but negligibly so. For an even rougher approximation of aircraft position, the lines do not need to be actually drawn on the chart, but can be visualized. This is often sufficient for a quick double-check of aircraft position.

With two VOR receivers this process is especially simple, since one can be used for primary navigation, while the other is used to provide more or less continuous cross-checks. With only one VOR receiver, however, it is necessary to retune the VOR to the second station. This means that the first radial will have to be noted and the last heading held until the second radial has been noted. This is only slightly more difficult than tuning the VORs independently, and the result is essentially the same.

When operating on the Victor Airways, many intersections are provided right on the charts. For instance, on Victor 16 (refer back to Fig. 2-9), Cream intersection has been plotted directly on the chart. One half of this intersection is the airway, and the other can be determined using a plotter and the compass rose around the intersecting VOR (Hampton). As you approach the intersection along Victor 16, with the intersecting radial set on the second VOR CDI, the needle on the second CDI will start to center, and when it has centered, you are at that intersection. If an intersection has been set up properly with a From indication on the intersecting CDI, the CDI needle will be on the side toward the intersecting VOR station before reaching the intersection, dead center at the intersection, and away from the station when past the intersection. This rule can be helpful in determining whether you have gone by an intersection or not.

The most accurate intersections are those that are defined by a nearby VOR at a 90-degree angle to the primary course. The further the intersecting VOR is away, the less specific the intersection defined by that radial will be. (Remember that radials fan out, and the "slice of pie" defined by one degree is much bigger away from the station than close in.) The more oblique the angle, the harder it will be to identify the exact moment the needle is centered; oblique angles create a larger area for the intersection. (See Fig. 2-13.) Neither of these points is critical; an intersection can still be defined using a distant VOR at an oblique angle, but the accuracy of intersections does vary, and this is why.

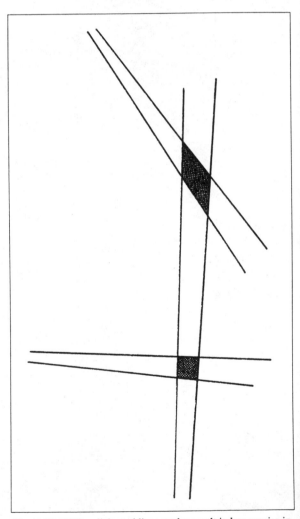

Fig. 2-13. VOR radials at oblique angles result in less precise intersection fixes than do radials at right angles.

TESTING

Regular testing of the VOR receiver and CDI is essential to safe operation. To operate under Instrument Flight Rules, a VOR test must be performed and recorded in the aircraft log or other record every 30 days. Testing is not required under Visual Flight Rules, but to the extent the VFR pilot relies upon VOR guidance, he would be well advised to do the same.

VOT Check

VOT stands for "VOR test facility". VOTs are installed at various major airports and are very easy to use. (The FAA lists these facilities in its *Airport/Facility Directory*; Jeppesen notes VOT facilities on the back side of the first approach plate for those airports so equipped.) To test a VOR receiver and indicator, the VOT frequency is tuned and the needle centered. The indicated radial should be within four degrees of either 180 degrees with a From indication or 360 degrees with a To indication to be acceptable for instrument use. This is also a good guideline for VFR use. Any deviation greater than four degrees indicates a problem in calibration or reception and should be resolved before further use.

Certified Check Points

Unfortunately, VOTs are normally only found at major airports; therefore, alternate testing procedures have been established. These consist of various ground and airborne check points that have been certified by the FAA for use as VOR receiver check points. To accomplish this test, the aircraft is positioned over or on the designated checkpoint, the appropriate VOR is tuned in and the needle centered, and the result is compared with the published standards. (These checkpoints are also listed in the FAA *Airport/Facility Directory*.)

The allowable tolerances for a certified checkpoint are plus or minus four degrees for ground checks and plus or minus six degrees for airborne checks. (Ground checks are preferable to airborne checks.) You will sometimes see signs on airport taxiways designating ground check points and listing the VOR frequency and radial to use for the check. This is a good way to test VOR receivers; it only takes a second, and while it is not as accurate as a VOT check, it will still reveal most malfunctions and calibration errors.

The FAA also allows you to test your VOR receivers by creating your own airborne checkpoint using the centerline of an airway radial and a prominent terrain feature. The exact procedure is spelled out in detail in FAR 91.25. I know of no one who has ever done this, but it can be done, in theory.

Comparison Check

An alternate method of VOR testing is to test two VOR receivers against each other. To accomplish this check, the two receivers are each tuned to a nearby VOR, the needles are centered and the radials noted. They should be within four degrees of each other to be acceptable. This is by far the most common method of VOR testing, since virtually all aircraft operated IFR are equipped with dual VOR receivers. It is not the most desirable though, since each system could be out of tolerance by an equal amount, and comparing the two would fail to detect this error. Still, this test will catch most malfunctions, and is much better than no check at all, particularly if it is regularly supplemented by a VOT check.

Calibration Checks

VOR receivers need to be tuned-up from time to time, just as engines do. The FAA recommends annual check-ups (the *Airman's Information Manual*, paragraph 4.a.), and states,

> If a [VOR] receiver's Automatic Gain Control or modulating circuit deteriorates, it is possible for it to display acceptable accuracy and sensitivity close into the VOR or VOT and display out-of-tolerance readings when located at greater distances where weaker signal areas exist. The likelihood of this deterioration varies between receivers, and is generally considered a function of time. The best assurance of having an accurate receiver is periodic calibration.

In other words, none of these checks is a substitute for a good, thorough check-out and tune-up by an authorized repair facility.

Reasonableness Testing

Even if the appropriate tests and check-ups have been recently accomplished, the accuracy and reasonableness of the VOR information should still be continuously verified in flight. If, for instance, the outbound and inbound radials on an airway do not match up well after a course changeover, then the set is probably out of calibration. If a needle stays centered for a long period of time without the need for any correction, then it probably is not working—it is unreasonable to expect a VOR needle to stay perfectly centered all the time. If a correction towards the needle results in the needle moving even further off center, check the orientation—it is probably backwards. Anytime the VOR guidance seems unreasonable, look for obvious errors. If none are found, attempt to verify your position with another source of information—a radar fix, a terrain feature, even dead reckoning. If your estimated position cannot be positively verified, then something is probably wrong, and you should retrace your steps, looking for errors that may not be obvious. Sometimes simply switching to another facility will correct the situation by forcing you to see a problem, such as an incorrect frequency, a reversed orientation, or even an Off flag not noticed before. Anyone can make a mistake and any well trained pilot can correct a mistake, but only an alert pilot can catch mistakes before they become a problem.

LIMITATIONS

The VOR system is an extremely capable system, but it does have certain limitations which must be observed. These limitations are discussed in the following paragraphs:

Range

The VOR system operates in the 108.0 to 117.95 MHz range. This range was deliberately chosen to provide clear, stable signals free from atmospheric effects. However signals in this range travel only in a line-of-sight, which limits their range to between 25 and 200 nautical miles (depending on aircraft altitude). This limitation means that long-range VOR navigation is possible only by linking a network of stations together, and it eliminates VOR as a practical, over-water navigational system (except for coastal areas and the local area around islands).

Terrain

VORs are fairly sensitive to terrain interference. While the FAA attempts to locate VOR transmitters on clear, elevated sites, this isn't always possible, and VORs will frequently be found to have limited range or erratic indications in certain directions, and some even have completely unusable quadrants. The Victor Airways are designed to avoid these unusable or limited quadrants, but it can be a limitation when operating VFR or off-airways. (VOR irregularities are published by the FAA and Jeppesen as a normal part of the en route instrument and approach chart service. VFR pilots should refer to the *Airport/Facilities Directory.*)

Propeller Modulation

Certain propeller (or rotor) speeds will cause deviations of up to 6 degrees in some VOR receivers. While this is not common, changing the propeller or rotor RPMs by even a small amount will usually correct the error.

Accuracy

The VOR system is limited to about one degree of accuracy. One degree at 200 miles represents a width of 3-1/2 miles. With an allowable error of plus or minus four degrees, this adds up to a total possible error of about 28 miles at a range of 200 miles. Therefore, if you were attempting to fix your location using two VORs, each at a range of 200 miles, your position could be anywhere within an area approximately 28 miles square. By comparison, long-range systems such as LORAN-C and OMEGA/VLF are capable of total errors of two miles or less, and over much greater distances. (But they have their own limitations.)

This concludes our discussion of the fundamentals of VOR navigation. The next chapter looks at another very useful piece of navigational equipment, DME (Distance Measuring Equipment). With these two chapters on VOR and DME fundamentals as background, the practical aspects of everyday VFR and IFR navigation are then examined in some detail in Chapter Four, VOR/DME Navigation.

Chapter 3

Distance Measuring Equipment

A VOR by itself provides only azimuth, or directional, information. Azimuth is normally sufficient for basic navigation—getting from here to there—but a single azimuth by itself doesn't tell the pilot where he is at any given moment. While, in theory at least, it doesn't matter if you don't know where you are, so long as you do know where you're going, in practice, position information is very important. This is particularly true in instrument conditions where aircraft separation is sometimes predicated on accurate positioning, but knowing where you are at some regular interval is also important in visual conditions—there is no other way to accurately monitor the progress of the flight.

Without distance information, position determination can be accomplished by manual methods: plotting position by the intersection of two or more LOPs (discussed in Chapter 2), or by triangulation, an older, geometric technique of little practical application to modern air navigation. (If you are curious about triangulation, see the *Instrument Flying Handbook*, Chapter VIII, 1980 edition, published by the FAA as Advisory Circular 61-27C, and available

through the Government Printing Office, Appendix A.) Manual methods of positioning are time consuming, are relatively inaccurate, are not continuous, and the results vary with pilot skill and the navigational aids (navaids) selected. Neither method provides ground speed information directly, and the entire process must be repeated each time a new position fix is desired.

As a result of these shortcomings, distance measuring equipment (DME) was developed to add continuous and highly accurate position information to the basic azimuth information provided by VOR. DME is not required for instrument or airway navigation, except for operations above Flight Level (FL) 240 (24,000 feet pressure altitude), and many pilots routinely and safely operate without DME, compensating for the lack of distance information with frequent cross radial checks. Nonetheless, VOR and DME should be thought of as essentially equal partners in an integrated system of navigation, VOR providing primary directional information, and DME providing primary position information.

DME is most often used by pilots operating under Instrument Flight Rules; under Visual Flight Rules, pilots can establish position by simply looking out the window. Having accurate and continuous distance information is therefore much less important to the VFR pilot than it is to the IFR pilot. Nonetheless, the security and safety inherent in having positive position and ground speed information has caused many VFR pilots to also invest in this not inexpensive, but still worthwhile, piece of equipment.

CO-LOCATION OF VOR AND DME

While DME is a completely separate and independent system from VOR, the FAA policy of *co-location*—pairing VOR and DME facilities at a single location—creates, in the FAA's words, "a unified navigational system." (AIM, para 6.a.) By co-locating VOR and DME facilities, the distance measured by the DME is the distance not just to the DME itself, but also to the VOR with which it is co-located. For all practical purposes, a co-located VOR/DME is a VOR that is capable of providing both azimuth and distance information—something VOR cannot do by itself.

VORs and DMEs, for technical reasons, operate on different frequencies: VOR between 108.00 MHz and 117.95 MHz, which is in the Very High Frequency (VHF) range, and DME between 962 MHz and 1,213 MHz, which is in the Ultra High Frequency (UHF) range. The two are integrated through the use of "paired frequencies." Paired frequencies means that every VHF VOR frequency has a standard UHF DME frequency assigned to it; however only one frequency, the VOR frequency, has to be selected in the cockpit—the paired DME frequency

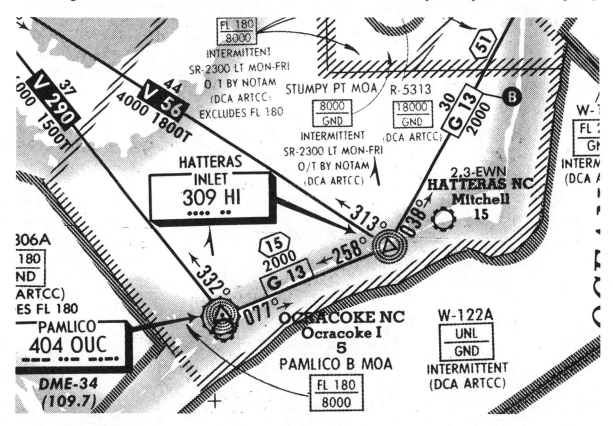

Fig. 3-1. The PAMLICO DME, indicated by the scalloped ring and the label "DME-34 (109.7)", can be channeled to the DME by tuning 109.7 MHz, but only distance information will be provided. G-13 is one of the very few remaining non-Victor Airways—an airway not based on a VOR facility—in the Contiguous United States. Reproduced with permission of Jeppesen Sanderson, Inc. Not actual size. Not to be used for navigation.

will be automatically selected as a function of the VOR frequency. To the pilot, the effect is the same as if the VOR and DME were the same facility with the same frequency. By dialing in the appropriate VOR frequency, both azimuth (radial from the station), and distance will be displayed; the combination of the two provides present position at any given moment.

DMEs do not have to be co-located. DMEs sometimes exist by themselves (see Fig. 3-1), in which case dialing the paired VOR frequency provides only distance to the station—the VOR CDI will show an Off flag. (Figure 3-1 is based on a Jeppesen Sanderson, Inc. instrument navigation chart, as are most of the instrument chart examples in this book. Jeppesen Sanderson, Inc.—commonly referred to as "Jeppesen" or "Jepp" for short—is the main commercial supplier of instrument navigation charts in the United States.)

In other cases, mainly on or around military or joint-use airports (both civil and military), DME and VOR are not co-located. (See Fig. 3-2.) Both VOR and DME information is provided, but the two facil-ities are separated by some small distance. In this case, the distance shown is not to the VOR, but to the DME facility (actually to the TACAN, which is a military VOR/DME—more on this later). These instances are not common, and will become even less so as the military completes its program of co-locating its DMEs with its VORs, but the possibility of a discrepancy in distance between VOR and DME should be noted whenever navigating to a VOR on a military or joint-use facility. These minor exceptions aside, DME means "Distance to the VOR."

Not all VORs have a DME associated with them, but the vast majority do. Those that don't are generally low-power, terminal VORs, intended mainly as approach aids to individual airports, or secondary Airway VORs. The VOR chart symbol will indicate whether DME is provided or not. Jeppesen indicates DME by a scalloped ring. (Refer back to the two previous illustrations.) If co-located, the DME ring will be centered on the VOR symbol. NOS charts indicate DME by additions to the basic VOR symbol. The Nantucket VORTAC, illustrated in the last chapter, Fig. 2-12, is one example of a VOR with DME

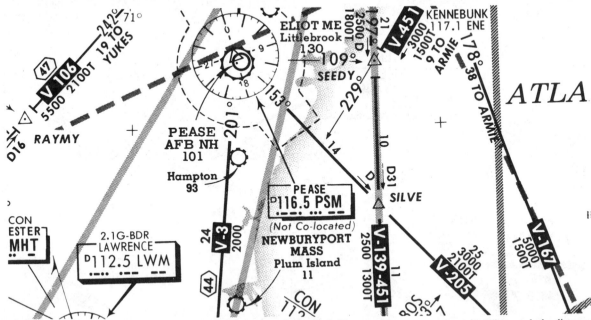

Fig. 3-2. The VOR and DME at Pease Air Force Base are available for civilian use, but are not co-located. When used, the distance shown on the DME will be the distance to the DME itself, represented by the dot inside the scalloped circle, and not the VOR (the dot inside the triangle)—a difference of about 1/2 nautical mile in this case. Reproduced with permission of Jeppesen Sanderson, Inc. Not actual size. Not to be used for navigation.

as depicted on an NOS chart. You will have to look fairly hard to find a VOR in the United States that does not have DME.

VOR/DME, VORTAC, AND TACAN

There is only one DME system, but it comes in three different forms: VOR/DME, VORTAC, and TACAN. The differences are minor, but to avoid confusion they should be explained.

The source of the confusion is the existence of two different VOR systems, both using the same DME system. The military VOR system is called TACAN. It was developed as result of what the military felt were shortcomings in the VOR system: terrain-induced errors, lack of portability, and the need for a level platform, negating carrier-based VORs. In addition, the military wanted DME to be an automatic, integral part of the system, not an add-on. The result is the TACAN system, which is a military VOR with automatic DME.

While these two nav systems—VOR and TACAN—are completely separate, the network of airways formed by them is the same for both the military and for civil aviation. This was done by locating VORs and TACANs in the same place—physically placing them side-by-side to create a common Airway fix. The FAA calls these combined facilities VORTACs. A VORTAC is nothing more than a combined VOR/TACAN facility.

But that's not the end of the story, because the two systems are not completely separate and incompatible. The azimuth portions of each are indeed separate and incompatible, but the DME portions are not: civil aviation can use the DME portion of TACAN. So a VORTAC, to a civilian pilot, is the same as a VOR with DME. (A VORTAC is the same as a TACAN with VOR to a military pilot.)

So what is a VOR/DME? A VOR/DME is a civilian VOR with DME and that's all—no TACAN. A military pilot can't use a VOR/DME (unless he happens to have a VOR receiver installed), but to a civilian pilot, a VORTAC and a VOR/DME are exactly the same thing.

If this is a little confusing, remember that *to a civilian there is absolutely no difference between a VOR/DME and a VORTAC,* but a TACAN by itself (the Pamlico TACAN, Fig. 3-1, for instance) can only provide distance information.

In this book, unless noted otherwise, when we talk about VOR/DME we also mean VORTAC, and when we talk about DME we also mean the DME part of TACAN.

PRINCIPLES OF OPERATION

Ironically, while DME is a very simple system in operation, it is a fairly complex system electronically. It is not absolutely essential that you understand how DME works, but it is very helpful in understanding the limitations and characteristics of DME operation if you do.

Active System

VOR is a passive system: signals move in one direction only, from ground to aircraft, and they do so continuously and without the need for any outside stimulus. DME, on the other hand, is an active system. It requires a transmitter and receiver on each end, and it requires an initiating signal from the airborne unit to begin functioning—it has to be "activated." In the process, DME signals are passed back and forth between the airborne unit and the ground unit, creating potential problems with timing and interference—problems which the two units must work out in cooperation to function properly. All of this makes DME a more complex system than VOR.

Pulse Cycle

DME operates on a pulse cycle. The airborne unit initiates the cycle by transmitting a pulse of UHF energy. As the pulse is initiated, a very accurate stopwatch starts. The pulse is called an interrogating signal because it "interrogates," or asks, the ground station for a reply. The ground station responds to this interrogation by transmitting a corresponding pulse back to the airborne unit. (The DME ground unit is actually a type of transponder, which is a contraction for transmitter-responder.) When the reply gets back to the airborne unit, the clock stops.

The round trip from airborne DME to the transponder and back is a cycle, and the amount of time it takes determines how far the two are apart: the longer it takes for a round trip, the further apart the two are. (Specifically, 12 microseconds—12/1,000,000 seconds—elapsed time equates to one nautical mile in distance.) The airborne unit looks at the amount of time involved in the cycle, translates that into nautical miles, and displays the result to the pilot: microseconds divided by 12 equals distance in nautical miles.

Sorting

If only one DME-equipped aircraft were to interrogate the transponder at a time, that would be the end of it. But of course there normally are many DME equipped aircraft within range of any given DME, all of them interrogating the ground station at regular intervals, all of them at different distances, and each demanding an individual response. The ground unit has to respond to each interrogation with a reply pulse, but the airborne unit is only interested in the replies that pertain to its interrogations. Some technique for sorting and identifying the replies as they are received by the airborne unit is necessary if the correct distance is to be displayed.

To do this, the airborne unit actually interrogates with not one, but with a pair of pulses. These two pulses are spaced in an irregular, random pattern that is unique to that unit. The ground unit replies to an interrogation with a pair of pulses that match the spacing of those received. The airborne unit then scans all the replies looking for a pair that matches its own interrogating pattern. When it finds one, it concentrates on that reply, and ignores the rest.

This process takes a finite period of time—the DME has to "watch" the replies for several seconds until it is sure it has identified its own pattern, then make the calculation and display the result. For this reason, when the DME is first tuned to a station it may be several seconds before any information is displayed, or the information may jump around before settling on a stable value. This isn't warm-up time (all DMEs manufactured in the last decade are transistorized and require no warm-up), it is search and identify time.

Lock-On

Once the pattern has been identified, we say the

DME is locked-on, and no further searching is necessary (unless the signal is interrupted for more than 10 seconds). Once the unit has locked-on, the distance display will be continuously updated, generally every tenth of a mile.

ACCURACY

Accuracy varies with the individual aircraft DME unit. The least accurate units are accurate to within three percent of the total distance: six miles at a normal maximum range of 200 nautical miles. Most units are more accurate than this, at least as accurate as 0.2 nautical miles at all distances, and some are capable of the maximum system accuracy of one tenth of a nautical mile at all distances. No other existing navigational system is capable of this degree of accuracy.

GENERAL OPERATION

DME may be the easiest piece of equipment in the cockpit to use: with one or two very minor exceptions, all you have to do is turn it on. Basic models do require that the VOR/DME frequency be manually set in the unit; i.e., in addition to setting the VOR frequency in the VOR receiver, the same frequency must also be set in the DME unit. This is an additional step, but it keeps the cost down, and it has the advantage of allowing the pilot to tune and receive a different DME from the VOR he is using, if he wants to. (It also allows him to forget that the distance displayed is not to the primary VOR, but to another VOR, if he isn't careful. More on this pitfall when we get to the Hold function.) Figure 3-3 is a picture of a typical basic unit requiring manual tuning. Notice that it has a frequency selector very similar to a VOR receiver's.

Most DMEs provide for automatic tuning. With automatic tuning, the DME will display the distance to the selected VOR without the need for any additional tuning—the VOR frequency is channeled directly to the DME. If more than one VOR is installed (usually the case), the pilot must select the VOR for which distance information is desired. Figure 3-4 is a picture of a unit with auto-tuning. Note the selector for Nav 1, Hold, and Nav 2.

The DME is normally channeled to the #1 VOR,

Fig. 3-3. A typical DME requiring manual tuning of the paired VOR frequency. Photo courtesy of Narco Avionics.

since that is the VOR most often used for primary navigation, however distance to the #2 VOR can be selected by throwing the switch. A typical situation where switching might be desirable is when navigating to a VOR near the destination airport, but with another VOR/DME actually located on the airport. With the #2 VOR set to the airport VOR, momentarily switching the DME from Nav 1 to Nav 2 will show distance to the airport, which is helpful in descent planning; distance to the primary VOR can then be reobtained by going back to the Nav 1 position.

Hold is used to retain a distance read-out when the VOR must be reset to a facility without DME. In this way distance information is still available, but from the previous facility, rather than through the fa-

Fig. 3-4. Most DMEs have provisions for auto-tuning of the DME frequency via the associated VOR frequency. When this is the case, a separate DME frequency selector is not required.

cility being used. Typically this is done when the VOR must be retuned from a VOR on or near the airport to an Instrument Landing System (ILS) frequency for an instrument approach. (ILSs normally do not have DMEs associated with them, although there are exceptions, and the numbers are increasing. See Chapter 8.) When Hold is selected, the DME is held on its last frequency and continues to provide distance information when it would otherwise be lost.

The problem with this procedure is that it is very easy to forget which distance is being displayed. It is critically important that you know which distance is being displayed at all times: VOR #1, VOR #2, a previously held frequency, or, in the case of a manually tuned unit, perhaps a completely unrelated VOR. If the distance displayed does not seem reasonable, it is probably because you are not looking at the DME you think you are—check the frequency or VOR selector to see which distance is actually being displayed. Several accidents have reportedly occurred because the pilot forgot, or was unaware, that the DME had been switched to Hold, assumed that the distance displayed was to the current fix and not the last one, and descended or turned too soon. The selector switch was put there to add versatility to the system, but in so doing the possibility for error is increased. Good operating practice is to check the selector switch (or frequency) each time the distance is consulted.

DERIVATIVE INFORMATION

DME provides only distance information directly, but it derives two important bits of additional data from that one piece of information: ground speed and time-to-station.

Ground Speed

The rate at which the distance changes is directly proportional to ground speed; all the DME has to do to get ground speed is to time the interval between distance changes, and compute the result. This is a basic dead reckoning procedure, similar to measuring the elapsed time between two known fixes and calculating the ground speed with an E-6B navigation computer. The differences between how DME figures ground speed, and the way it is done manually are: (1) DME can use very small distance intervals (tenths of a mile, where you might want to use five or 10 miles to be accurate); (2) DME uses an internal computer rather than a mechanical computer; (3) DME can compute ground speed continuously and provide regular updating of the ground speed read-out. Frequent updating is especially helpful in noting changes in wind direction or velocity from one area to another, or from one altitude to another.

To avoid errors in interpreting ground speed read-outs, it is important to remember the way in which DME calculates ground speed: it looks at changes in the distance, and measures the elapsed time between those changes to calculate ground speed. This works fine most of the time; i.e., when the aircraft is going directly to or away from the station. But if the aircraft is flying by the station, or any other relationship to the station other than tracking straight at it or straight away from it, the distances will no longer change in direct proportion to the ground speed of the aircraft and the ground speed read-out will no longer be meaningful.

The easiest way to see this is to consider the situation where an aircraft is flying in a circle around a DME. The DME distance will be constant, because the aircraft is flying in a circle about the station; therefore there will be no change in the distance to or from the station. Since the distance doesn't change, as far as the DME is concerned the aircraft must be standing still and the DME will read a ground speed of zero. The same thing happens flying by the station, only in this case there is some movement to or away from the station, but less than the actual ground speed.

The DME is right, in its own way. The ground speed to the station is zero when flying in a circle about it, but that isn't the information you are looking for when you select Ground speed (more commonly Knots). This is one of those areas where a label doesn't always mean what it says. The label should probably read "Ground speed to or from the Station" to be complete and accurate, but that would take up a lot of space on the indicator. Just remember that the ground speed read-out is valid only when tracking directly to or from the station.

Time-to-Station

Once you have distance and ground speed, time-to-station (in minutes) is just one more calculation. This is another basic dead reckoning problem and something the internal computer can handle very easily. It is also something the pilot could handle very easily with an E-6B—ground speed over the arrow, time under the distance—but it is so easy to add this feature that most DME units include it. The DME illustrated in Fig. 3-4 allows for the selection of either ground speed or time-to-station for display in the bottom window. Selection is done with the knob marked "Kts Min Off."

IDENTIFICATION

Since the DME is normally tuned to the same frequency as the VOR (often automatically), as long as the VOR is itself identified, it generally isn't necessary to also identify the DME. If the proper VOR is being received, then it can be assumed that the proper paired DME frequency is also being received. The DME does, however, have its own identification signal which can be used to verify both tuning and proper operation of the ground based DME unit if desired or necessary. The DME ident code is the same as its corresponding VOR ident code, but it is transmitted at a slightly higher pitch and at less frequent intervals (approximately every 30 seconds).

If the DME seems to display an unreasonable distance, possibly indicating a tuning problem, the coded ident will indicate the station actually being received. In the event of a lock-on failure, listening for a DME ident will tell you if the ground unit is functioning.

CAPACITY

The DME ground unit can only handle a finite number of interrogations from different aircraft before becoming overloaded. Overloading is a rare occurrence, and is generally confined to major VOR/DMEs during times of heavy aircraft movement. Most DMEs can handle all of the traffic they need to, but overloading is something you should be aware of, since it may account for an otherwise unexplainable, temporary malfunction.

An overloaded DME ground facility continues to reply to its original "customers," but it won't accept any new ones until one of the current users goes off the frequency. To a pilot, an overloaded DME acts like one that is out-of-service—it won't lock-on. It is a common courtesy during busy times at major airports to turn the DME off while on the ground in order to insure service to airborne aircraft.

SLANT RANGE ERROR

DME measures the distance in a straight line from the aircraft to the ground unit. DME distance is called Slant Range, since the distance measured is not along the ground, but on a line that slants from aircraft to ground. It introduces a small but unavoidable error called (logically enough) Slant Range Error. (See Fig. 3-5.) Actually, slant range is not an error at all—slant range represents the true distance from aircraft to ground unit—however, the pilot is not interested in true, straight line distance between aircraft and ground, but in the distance between his position and the DME, measured over the ground. Because of this discrepancy, it appears to be an error.

The closer the aircraft is to the station, and the higher the aircraft altitude above the station, the greater the difference between slant range and ground distance. Far away and low, the difference between slant range and ground distance will be minimal, however, close in and high the difference will be great. When the aircraft is directly over the station, the distance shown will be the aircraft height above the DME in nautical miles. At an absolute altitude of 41,000 feet, for instance, the DME will indicate 6.7 nautical miles when directly overhead, even though the actual distance-to-the-station over the ground at that point is zero.

As a rule of thumb, slant range error can be disregarded when the aircraft is more than a mile away from the station for every thousand feet of altitude above ground level; i.e., more than five miles away at an altitude of 5,000 feet above ground level (AGL), 35 miles away at FL 350, and so on. (At 5000 feet AGL and five nautical miles, the slant range error is less than 0.1 nautical miles, and at FL 410 and 41 nautical miles, the slant range error is about 0.6 nautical miles.)

Slant range error also affects ground speed ac-

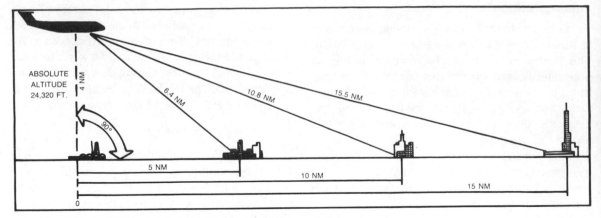

Fig. 3-5. DME measures slant range, which is not the same as ground range.

curacy. As the aircraft nears the station, the slant range error varies from nil to fairly substantial. For instance, at 5000 feet AGL and 5.0 miles, slant range is 5.1 nautical miles, but the error increases rapidly as the aircraft nears the station to 0.8 nautical miles slant range when directly over the VOR/DME. In effect, "extra mileage," at an increasing rate, is added on as the aircraft approaches the station, and as a result the aircraft appears—to the DME—to slow down somewhat. Then, as the aircraft passes over and beyond the station and slant range error diminishes, the aircraft appears to speed up again. The ground speed will again stabilize on an accurate value when the airplane is far enough away from the station that slant range error is again negligible.

It is not imperative that you understand exactly how slant range error affects ground speed, but it is important that you understand that it does, and that you not base any calculations, such as time-of-arrival or fuel required, on ground speeds taken when approaching or departing the station. (Approaching or departing means the area where slant range error is significant: one mile for each 1,000 feet of cruising altitude above ground level.) If you do, the ground speed will be slow by an amount that is directly proportional to the slant range error, and your estimates will be wrong by that factor.

DME ARCS

In addition to providing distance, ground speed, and time-to-station, DMEs have certain specialized navigational uses. One of these is the DME arc.

A DME arc is flown by turning the aircraft just enough to maintain a constant distance from a given DME. DME arcs are used in air navigation in a manner that is similar to transitioning from a major highway to a beltway around a city. The airway represents the major highway, and the DME arc serves as the beltway, taking the aircraft from the en route course to the inbound approach course. This will be covered in greater detail in Chapter 8, Instrument Approaches.

The specific techniques for flying DME arcs are best learned in the aircraft or simulator. (They are not as easy to do as they look, especially without an RMI, because wind is a big factor; a DME arc is a partial turn-about-a-point done solely by reference to instruments.) For our purposes, it is sufficient to note that distance information can be used to navigate along an arc; its uses are limited, but not unimportant.

RHO-RHO NAVIGATION

The navigational symbol for distance is the Greek letter rho (written either P or ϱ). The corresponding symbol for azimuth is theta (written Θ or θ.) Most navigation is either theta-theta based or theta-rho. For instance, using the intersection of two VOR derived LOPs to determine position is theta-theta navigation, while VOR/DME position is theta-rho navigation. Rho-rho navigation uses distance exclusively.

Rho-rho navigation is quite a bit more complicated than either theta-theta or theta-rho, and requires an onboard computer to do all but the easiest calculations. It is, however, very accurate—much more accurate than any system which uses theta, for the simple reason that DME is so much more accurate than any other nav system.

While the practical applications of rho-rho navigation are complicated and sophisticated (jump ahead to Chapter 17, Navigation Management Systems, if you want to know more at this point), the concept is not too difficult, and understanding how a rho-rho system of navigation works will give you a good idea of the versatility inherent in the DME system.

Just as azimuth alone (radial from the station) says that your present position is somewhere along a line, distance alone says that your present position is somewhere on a circle (after being corrected for slant range error). If, for instance, you know that you are 10 nautical miles from a DME (and that is all you know), you know you could be anywhere around that DME on a circle with a radius of 10 nautical miles. If you also know your distance from another DME, then you know you are also somewhere around a circle of that size from that DME. Where the two circles cross each other is your present position.

Unfortunately, unlike the intersection of two lines, unless the circles happen to be exactly tangent, the intersection of two circles creates not one point, but two. (See Fig. 3-6.) It takes a third circle to determine which of the two is the correct one. Once

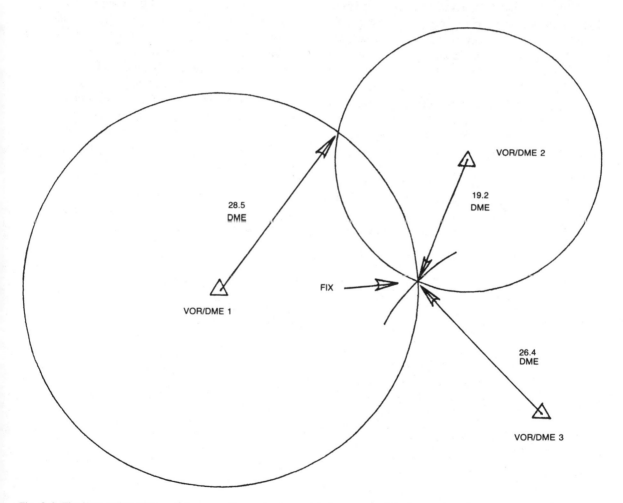

Fig. 3-6. The intersection of two circles normally creates two points in common. The intersection of a third circle creates a fix.

the third circle has identified the correct intersection, that DME is no longer needed (which is why the system is called rho-rho and not rho-rho-rho navigation—thank goodness for small favors).

This part of rho-rho navigation—position finding—is not too difficult. You could, if both of your VORs were inoperative for instance, take a couple of different DME readings, and using your plotter as a crude compass, draw arcs on your chart from each of the DMEs selected to see where the circles intersected. Then a quick check of distance from a third DME (the third arc would cross only one of the two intersections) would indicate which of the two intersections was the correct one.

Without a real compass to draw the arcs (something pilots don't normally carry with them), this would not be very accurate, but it would be better than nothing in a pinch. Once your present position was known, you could then take up a heading in the desired direction, and by making repeated position checks with the DME, you could correct the heading as necessary to proceed, in a somewhat wobbly fashion, to your destination. The more frequently distance checks were made, the more accurate the course would be.

In practice, the only way to navigate with rho-rho navigation in anything resembling a straight line is to have an onboard computer to do the chartwork. A computer can take the continuous distance information provided by several different DMEs to keep a running tally of your progress along the desired course, and then tell you which way and how much to turn to get back on course. The main advantage to a rho-rho navigational system is that the course described will be much more accurate than that described by VOR alone, which decreases the distance traveled and increases the efficiency, and the desired course can fall between any two random points— you are not limited to courses linked by VORs.

SUMMARY

DME is most commonly used as a supplement to VOR Airway navigation by providing continuous distance information, ground speed computations, and time-to-station estimates. This is its primary and original mission, and it does it well. But it is capable of doing much more, as I hope this brief discussion of DME arc and rho-rho navigation has shown.

The next chapter, VOR/DME Applications, moves away from background information, and into practical applications of VOR and DME navigation for VFR and IFR operations, both on and off airways. With the background information from these first three chapters as a reference, it should become clear in the next chapter why the VOR/DME-based airway system exists as it does, and why certain techniques for using it work better than others.

Chapter 4

VOR/DME Navigation

There are over a thousand VOR stations in the Continental United States (and many more around the world), and almost as many DME facilities. These facilities form a network of stations—a network that is capable of providing complete navigational guidance from any one point in the U.S. to any other at all altitudes from 1,000 feet AGL to FL 450. This network of stations can be used independently (mainly VFR), or as part of a system of airways for instrument navigation (the Victor Airway and Jet Route systems).

VOR/DME navigation by itself is not capable of providing navigational guidance from any random point to any other random point in a direct course—that is, in a straight line. The requirement to fly along radials extending out from the facility is a characteristic of VOR/DME navigation that cannot be avoided without additional equipment (see the next chapter, VOR/DME-Based RNAV); however, VOR coverage is so complete in the United States (and in most of the industrialized areas of the world as well), that deviations from the straight line distance are typically less than 10 percent when using VOR guidance,

and on longer trips the deviation is typically less than five percent. This is a very small penalty to pay for a system that is easy to use, relatively inexpensive, accurate, and very reliable.

VFR VOR NAVIGATION

VOR stations can be used directly for VFR navigation, without regard for the airway system, at the discretion of the pilot. That is, a pilot may use any VOR within range to navigate either directly To or From that VOR, as necessary to go from point A to B. A few examples will illustrate how this is done.

Direct Courses

To navigate from Lexington, Kentucky, (Blue Grass Airport) to Nashville, Tennessee, (Metro Airport) using VOR facilities directly, the pilot starts by drawing a line on his chart between the two airports. (See Fig. 4-1.) He then notes the nearest VORs along the direct route of flight, which in this case are: Lexington (LEX, 112.6 MHz), New Hope (EWO, 110.8 MHz), Bowling Green (BWG, 117.9 MHz), and

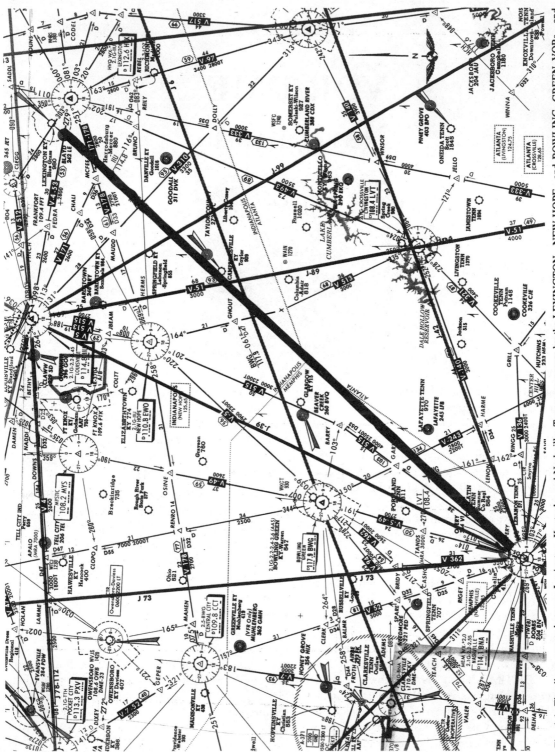

Fig. 4-1. The direct course between Lexington, Kentucky and Nashville, Tennessee passes by the LEXINGTON, NEW HOPE, and BOWLING GREEN VORs, and ends at the NASHVILLE VOR. (An instrument chart has been used here for clarity, but the process is exactly the same using Sectional or WAC charts.) Reproduced with permission of Jeppesen Sanderson, Inc. Not actual size. Not to be used for navigation.

Nashville (BNA, 114.1 MHz). Since the first VOR, Lexington, is quite close to the airport but considerably off to the side of a direct route of flight, the pilot would probably elect to disregard this VOR. The first VOR used would then be the New Hope VOR. After takeoff from Lexington, the pilot would initially take up a heading in the direction of New Hope. Once New Hope was within reception range and had been identified, he would center the needle and track it directly to the station. (Refer back to Chapter 2 for a review of specific techniques on tracking to and from VORs.)

Once over the New Hope VOR, he would retune the VOR to the Bowling Green VOR, re-center the needle, and begin tracking again. Once past BWG he would retune the VOR to Nashville, repeat the process, and then navigate directly to the destination: Nashville Metro Airport.

The straight line distance between Lexington and Nashville airports is 152 nautical miles, and the distance using EWO and BWG VORs is 57 plus 56 plus 50 nautical miles, for a total of 163 nautical miles—11 miles extra (or about seven percent). For most pilots, this is a fairly small price to pay for the convenience and security of being able to navigate directly from one VOR to another and avoid the uncertainties of pilotage and dead reckoning. Navigating in this way, always to the next VOR within reception range that is closest to the straight line course is easy, accurate, and requires the minimum in cockpit instrumentation: a single VOR receiver and CDI. It does, however, have its limitations.

Standard Service Volumes

VOR range is limited by the line-of-sight characteristic of VHF transmissions: reception distance is a direct function of aircraft altitude. The FAA has established standard reception distances for VORs according to station power and aircraft altitude above the station. It calls these reception distances Standard Service Volumes—an unusual choice of words, but volume in this case refers to the volume of space about the VOR—the area above as well as the area surrounding the VOR in which positive reception—service—is assured. Figure 4-2 illustrates the standard service volume for three different classes of VOR:

terminal, low altitude, and high altitude.

The VFR pilot is primarily concerned with the standard low altitude service volume—the figure in the upper right hand corner of Fig. 4-2. (Terminal VORs, represented by small VOR compass roses on the charts, are not commonly used for en route navigation. They are limited to 25 nautical miles range from 1,000 to 12,000 feet AGL, and are mostly used for instrument approaches.) The standard low altitude VOR is limited to 40 nautical miles in range for altitudes between 1,000 feet and 18,000 feet above station elevation.

The low altitude en route and sectional charts do not differentiate between low altitude and high altitude VORs, but this is of little consequence to the VFR pilot; the only difference between a high and a low altitude VOR below 18,000 feet (above which VFR flight is not allowed) is between 14,500 and 18,000 feet, where the reception distance for the high altitude VOR is increased from 40 to 100 nautical miles. Since few VFR pilots fly at these altitudes, this is a very minor exception. For all practical purposes then, the positive reception range for en route VORs is 40 nautical miles for VFR operations.

Referring back to our example above—Lexington to Nashville—we noted earlier that the distance to each of the VORs is greater than 40 nautical miles: 57 nautical miles to EWO, 56 nautical miles to BWG, and 50 nautical miles to BNA. Since climbing to at least 14,500 feet AGL to take advantage of the 100 nautical mile Standard Service Volume of High Altitude VORs between 14,500 and 18,000 feet is probably not practical, and since only BWG and BNA happen to be High Altitude VORs anyway (you would need to check either the *Airport/Facility Directory*, an FAA publication listed in Appendix A, or a high altitude en route chart to be sure of this), this means that in this case positive course guidance cannot be assured for the entire route of flight simply by navigating directly to each of the VORs.

For VFR flight this is not a serious limitation. There is no requirement that VFR pilots navigate using VORs at all—they are simply an aid to VFR navigation. On a VFR flight plan, pilotage and dead reckoning can be used for primary navigation, with VOR navigation used as a supplement as available.

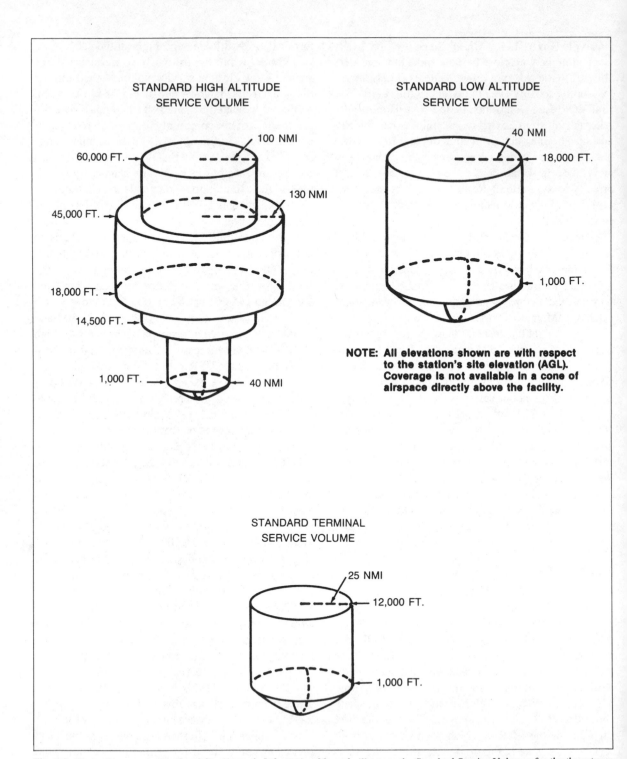

Fig. 4-2. These diagrams, taken from the *Airman's Information Manual*, illustrate the Standard Service Volumes for the three types of VOR facilities. Positive VOR guidance is available only within the Standard Service Volume.

Alternatively, pilotage and dead reckoning can be used initially to head in the general direction of the VOR, changing over to VOR navigation once the VOR comes into range. (This is the method used above.) This method of proceeding directly to a series of VORs, supplemented as necessary by dead reckoning, works, but it is less than optimum.

Linking VORs

Instead of always tracking to the next VOR until the destination (or the closest VOR to the destination) is reached, and relying on dead reckoning for directional guidance until within VOR reception range, the VOR itself can be used to track From the last VOR, in the direction of the next VOR.

For instance, on the leg between EWO and BWG (refer back to Fig. 4-1), note that a line has already been drawn on the chart between these two VORs, and the outbound radial is indicated: 220 degrees outbound from the EWO VOR takes you directly to the BWG VOR (and conversely, the 039-degree radial outbound from the BWG radial takes you directly to the EWO VOR). After tracking inbound to the EWO VOR, the pilot notes station passage, turns the OBS to 220, turns the aircraft in that direction, establishes himself on the outbound course, and then tracks outbound. When he feels he is within reception range of the BWG VOR, he retunes the VOR receiver to that frequency, identifies the BWG VOR, centers the needle with a To indication, and then proceeds as before directly to the BWG VOR. Crossing the BWG VOR, he notes the outbound course to BNA (191 degrees), resets the OBS, and repeats the process to BNA.

The advantage to this technique of using one VOR outbound to the next VOR inbound is that in so doing the effective range of positive VOR navigational coverage is doubled, from 40 to 80 nautical miles. Since none of the legs between VORs on this particular trip is separated by more than 80 nautical miles, this method will provide the pilot with positive navigational guidance from shortly after takeoff to touchdown.

These lines between VORs are actually airways (discussed in much greater detail in the next section), and essential airway data such as airway identification (V-5 here) and airway radial in degrees magnetic are marked on VFR charts as well as on IFR charts—the instrument chart is used here only for clarity.

VOR Navigation Outside Standard Service Volumes

In practice, it is not always necessary to limit VOR guidance to 40 nautical miles or less, or 80 nautical miles between VORs. At normal VFR cruising altitudes (3,500 to 9,500 feet MSL for instance), most VORs can, in fact, be received at distances greater than 40 nautical miles, but the navigational guidance will not be positive—that is, the FAA will not guarantee that the azimuth information provided beyond that distance will be free of interference or accurate to Airway standards for Instrument Flight Rules in all directions. Since the VFR pilot is not required to observe the limitations of Instrument Flight Rules (but neither is he fully able to take advantage of their protection), and since he does have visual reference to the surface (which should enable him to spot any inaccuracies or errors), the VFR pilot can frequently make use of VORs beyond their Standard Service Volume distances. To the extent the VFR pilot is relying upon VOR guidance for course information, however, he should still observe the standard service volume limitations. This means navigating directly to and from low altitude VORs that are no more than 80 nautical miles apart (200 miles apart for high altitude VORs between 14,500 and 18,000 feet AGL).

Destination Guidance

In the previous example, the last VOR used was the BNA VOR, conveniently located right on the Nashville Metro Airport. In this case, navigating directly to the BNA VOR also means navigating directly to the Nashville airport. But not all airports have VORs located directly on them, in fact the majority do not, and many do not even have a VOR near enough to insure being able to see the airport from that VOR. VOR information can still be used in these cases to help find the destination airport (or any other specific point for that matter) provided that the VOR to be used is within range of the airport (normally no more than 40 nautical miles away).

On the return trip, for instance, Nashville to Lexington, Lexington airport does not have a VOR located on the airport, but there is a VOR nearby—the Lexington VOR (Fig. 4-1). The radial From the VOR to the airport can be determined using a straight edge and the VOR compass rose printed on the chart. (Special plotters are also made for this purpose, and commercial publications are available which list, with considerable accuracy, radials and distances from nearby VORs for nearly all public airports.) In this case, the desired radial is the 305-degree radial from the LEX VOR. After navigating To the LEX VOR, the pilot waits for station passage, and just as he did when navigating outbound from one VOR to another, he resets the OBS outbound to 305 degrees, turns the aircraft in that direction, and then intercepts and tracks that radial directly to the destination airport. (A single VOR can't, by itself, tell the pilot when he is exactly over the airport, but it can take him on a course that crosses over it; to positively identify the fix itself would require either visual confirmation, distance measuring equipment, or a cross radial from another VOR.)

VOR Course Interception

One last way in which the navigation for this round trip from Lexington to Nashville and back using basic VOR guidance could have been improved upon involves the first leg from takeoff to the EWO VOR. There is a way to obtain VOR course guidance from the Lexington VOR, without having to fly out of the way directly to that VOR. This is called a VOR course interception, or course intercept for short.

There is an airway between the LEX VOR and the EWO VOR, and the outbound radial for that airway from the LEX VOR is 251 degrees. To intercept this radial, the pilot would tune in the LEX VOR and set the CDI to the 251-degree radial. Then, the pilot would estimate a heading after takeoff that would intercept this 251-degree radial—something like 210 degrees would be about right, which would be close to a 40-degree interception angle. (Anything between about 180 degrees and 230 degrees would be satisfactory. Anything less than that would take the aircraft backwards, to the east, and anything much greater than that would put the aircraft into the area

where the EWO VOR could be received anyway, negating most of the value of the exercise.)

Once a heading to intercept has been determined, all the pilot has to do is maintain that heading until the needle, set to 251, starts to center. As it does, he turns the aircraft in the direction of the outbound course (251 degrees in this case), and corrects towards the needle as necessary to complete the course intercept. Once established on course he can begin tracking, and as soon as he is within 40 nautical miles of the EWO VOR he can change over and track inbound to that VOR. By intercepting the outbound course of the LEX VOR, the pilot reduces the amount of time he must navigate without VOR guidance, without having to fly out of his way to do so.

With these basic techniques—flying directly To a series of VORs, linking VORs by alternately tracking inbound and outbound in as close an approximation of a straight line course as possible, navigating along selected radials to desired locations, and intercepting courses as necessary to minimize dead reckoning—the VFR pilot can take advantage of the extensive network of VOR facilities to virtually ensure safe, reliable, and accurate navigation anywhere in the Continental United States.

THE VICTOR AIRWAY SYSTEM

The Victor Airway System is a formal network of airway routes, based on the VOR/DME system, for altitudes beginning 1,000 feet above the surface and continuing on up to 18,000 feet above the surface. The Victor Airways are primarily an IFR system, but they are available for VFR navigation. The Victor Airway System is sometimes called the low altitude airway system, and the instrument charts which depict it are called low altitude en route charts.

The advantage to a formal airway system is that, so long as the airway limitations depicted on the charts are observed, the pilot does not have to concern himself with the limitations of the individual VORs and DMEs themselves. That is, he does not need to worry about terrain restrictions, standard service volumes (i.e., altitude and range), or frequency interference, nor does he need to be concerned with whether a particular VOR happens to be a low or a high altitude facility. The Airway system takes all of these factors into account in its design. In addition, the FAA regularly flight checks the airway system to ensure that the published restrictions are sufficient to protect the pilot from the specific limitations of the individual VORs and DMEs which comprise it.

The Victor Airway System also provides the pilot with many conveniences, such as a variety of regularly spaced intersections, predetermined radial information, distances between fixes, and in some cases, outbound/inbound changeover points. An airway system is inevitably more restrictive than a non-airway system, however the greater level of safety and convenience created generally out-weighs any disadvantage in efficiency.

VICTOR AIRWAY ROUTES

Victor Airways are linked in extended route systems, using numbered identifiers. Victor 3 (V-3), for instance, starts at Presque Isle, Maine (PQI VOR), and continues through a series of VORs down the East Coast, terminating at the Key West VOR (EYW, 113.5 MHz) at the tip of Florida. Victor 2 (V-2) starts at the Lawrence VOR (LWM, 112.5 MHz), near Boston, and continues all the way across the United States to the Seattle VOR (SEA, 116.8 MHz). (Perhaps my Northeast bias shows here: I could just as easily have said that V-3 starts in Florida and ends in Maine, and V-2 starts at Seattle and ends near Boston—with very few exceptions, airways work equally well in both directions. The exceptions are clearly marked on the charts by one-way arrows.)

Not all Victor Airways extend for such great distances, nor can a pilot always count on a single airway taking him from his departure point to his destination (any more than a single Interstate highway always will when he travels by car), but in most cases pilots can take advantage of the Victor Airway System over at least portions of their routes. To the extent a pilot is able to use the Victor Airway System, his flight planning, flight plan filing, clearance copying, and en route navigation will be substantially simplified.

VICTOR AIRWAY CHARACTERISTICS

The instrument chart used in Fig. 4-1 is part of the Jeppesen U.S. (LO) 28 low altitude en route chart. Since the majority of civil aviators use Jeppesen charts for instrument operations, they have been selected for purposes of illustration throughout this book. The markings used on officially produced National Oceanic Service (NOS) low altitude en route charts (Government charts) are similar to those on Jepps and provide identical information, but they differ in certain small details of presentation—those who prefer Government charts should have no trouble following this discussion. (Information on obtaining both Jeppesen and NOS charts is provided in Appendix B.) The Jeppesen U.S. (LO) 28 low altitude en route chart will be used for most of the remaining examples in this chapter.

Figure 4-3 shows a typical Victor Airway segment between Central City and New Hope VORs. This particular airway is called V-178 (dark box); it continues on to the west of Central City VOR and on to the east of New Hope (not shown)—in fact, V-178 connects the Hallsville VOR near Columbia, Missouri with the Bluefield VOR in Bluefield, West Virginia. This segment of V-178 is described by the 075-degree radial from the Central City VOR (109.8 MHz, CCT), and the 258-degree radial from the New Hope VOR (110.8 MHz, EWO). The minimum altitude required to ensure adequate VOR reception and terrain clearance between the two VORs is 2,700 feet MSL, indicated by the four-digit number under the airway marking. The distance between Central City and New Hope VORs is 77 nautical miles (indicated in the smaller box with pointed ends), and there are four individual route segments between the two VORs of 10, 22, 14 and 31 nautical miles each (77 total), as indicated by the numbers without boxes in the middle of each segment. The fixes that separate these segments have five letter names: Mahen, Renro, and Osine. (Intersection fixes have five letter names for compatibility with the FAA computer system and for standardization. This makes for odd sounding, and often difficult to pronounce, nonsense names, but the computer likes it and the standardization does help eliminate ambiguity between intersection names and VOR facilities.) Airway intersections can be used for position finding, position reporting, turning points, and transition from one airway to another. (V-49 in-

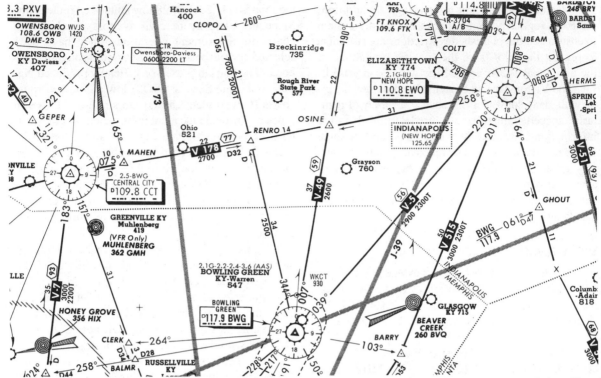

Fig. 4-3. A typical Victor Airway segment between CENTRAL CITY and NEW HOPE VORs. Reproduced with permission of Jeppesen Sanderson, Inc. Not actual size. Not to be used for navigation.

tersects V-178 at Osine intersection, for instance.) These are the basic characteristics and markings common to all airway segments.

FLYING THE VICTOR AIRWAYS

The procedure for getting established on a Victor Airway is similar to that for VFR en route navigation using VORs independently: initially, the pilot must either fly directly to a nearby VOR, or intercept an airway radial along the desired route of flight. Once established on an airway however, certain procedures must be observed—airway navigation provides many safeguards, but, as an inevitable consequence of those safeguards, the pilot is more restricted in his use of the system.

Airway Navigation—Outbound

The pilot is obligated to use the alternating outbound/inbound method of linking VORs, when navigating via Victor Airways. This means that he cannot jump from one VOR to another, always navigating To the next station, but must navigate From one, To the next. Therefore, departing the Central City VOR on V-178 (Fig. 4-3), the pilot selects the 075 radial with a From indication on his CDI and corrects as necessary to maintain a track on the centerline of the airway. (Quarter scale needle deflections or less will insure that the aircraft stays within the lateral boundaries of the airway.) He continues on this course outbound until time to change over to the inbound course to the next VOR along the route of flight.

Standard Changeover Points

Since radials expand in width as the distance from the VOR increases, the course guidance decreases in accuracy as the aircraft flies outbound. At Mahen intersection, for instance, a course width of one degree is about 0.2 nautical miles across; at Osine intersection the same course is 0.8 nautical miles across. Normally, the greatest course accuracy comes from flying outbound from one VOR to the halfway point, and then flying inbound on the corresponding radial to the next VOR—"changing over." In the absence of instructions to the contrary, the standard changeover point is therefore midway between any two VORs.

Since the route segment between CCT and EWO VORs is 77 nautical miles, the midway changeover point is 38.5 nautical miles, or about halfway between Renro and Osine intersections. It is not imperative that the changeover occur at the exact halfway point. The closer it is done to the halfway point, the more accurate the course guidance will be, but anywhere more or less halfway between Renro and Osine intersections would be acceptable.

If DME is installed, identifying the halfway point is a fairly easy process—divide the leg distance in half and change over when that distance appears on the DME. In the absence of DME, a published intersection close to the midpoint can be used, an unpublished midpoint intersection can be set up (the 350-degree radial off Bowling Green is very close to halfway between CCT and EWO VORs), or the midpoint can be estimated on the basis of time.

Airway Navigation—Inbound

To minimize chart clutter, en route charts, both NOS and Jeppesen, only show the outbound radials. In order to have proper needle sensing when tracking inbound, the pilot must select the reciprocal of the outbound, or published, radial. (*Reciprocal* means *opposite* here.) Therefore, inbound to EWO, the pilot selects not the 258-degree radial (which is the outbound course from EWO VOR for V-178), but the reciprocal of 258, which is 078 (258 minus 180 equals 078).

Most CDIs show the reciprocal to the course selected at the bottom of the indicator. Therefore, a simple way to select the inbound course is to put the outbound (published) course at the bottom of the CDI; the inbound course will then be automatically selected at the top. This avoids errors in adding or subtracting 180 to obtain the reciprocal.

Regardless of how the inbound course is determined, as long as the pilot is careful to always use the published airway radials outbound, and the reciprocal of the published radials inbound, the course orientation (the To/From flag) will always be correct. It never hurts of course, to double check the flag; a To flag outbound or a From flag inbound means you have it backwards.

MAGNETIC ALIGNMENT

Logically, the inbound course to the next VOR should always be 180 degrees opposite from the outbound course of the preceding VOR; that is, if the outbound course from one VOR is 090 (due East), you would expect the published radial for the inbound course to the next VOR to be 270 (due West)—the reciprocal of 090. But on V-178, the outbound course from Central City is 075 and the inbound course to New Hope is 258—not 255 as you would expect (the reciprocal of 075). The explanation for this lies in the fact that VORs are aligned with magnetic north, and as the variation changes, the magnetic alignment of the VORs also changes. Depending on the amount of change in variation between the two (VORs aligned North and South of each other will generally have less change in variation than those aligned East and West), the inbound and outbound courses will differ by more or less than exactly 180 degrees. This is another reason why it is important to change over at the appropriate point—the pilot who continues on the 075 degree radial off Central City all the way to New Hope will miss New Hope by almost 1 1/2 nautical miles due to the change in magnetic variation alone.

NON-STANDARD CHANGEOVER POINTS

While the halfway point between two VORs is the theoretical optimum changeover point, in those cases where terrain restricts one signal more than another the optimum changeover point may not occur exactly halfway between the two, necessitating a non-standard changeover point. Figure 4-4 shows an airway with a particularly unbalanced changeover point: V-140 between Nashville and Dyersburg VORs. Delha intersection is marked by a pair of L-shaped symbols; the symbol that points to the left has "92" inside it, and the symbol that points to the right has "36" inside it. The L-shaped symbols indicate a nonstandard changeover point for this segment of the airway, and the numbers mean that the exact distance

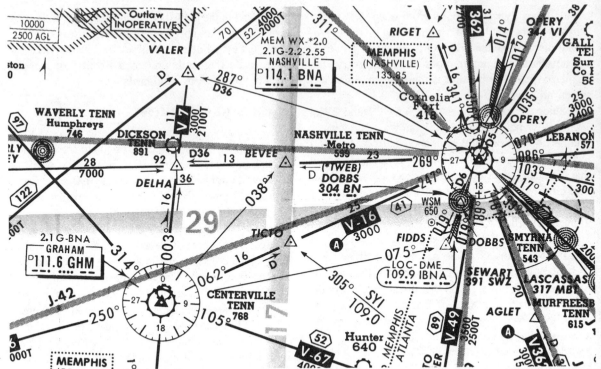

Fig. 4-4. A non-standard changeover point is indicated by the bracket shape symbols at DELHA intersection. The numbers inside the brackets (92 and 36) indicate DME distances for the changeover point. Reproduced with permission of Jeppesen Sanderson, Inc. Not actual size. Not to be used for navigation.

to change over is 92 nautical miles from the VOR to the left (Dyersburg), or 36 nautical miles from the VOR to the right (Nashville). (The 003 radial off Graham VOR could also be used to identify this intersection.) For some reason—probably terrain, but possibly frequency interference—FAA airway personnel have determined that the most accurate course guidance is provided by changing over to the next VOR at the indicated non-standard changeover point, rather than the standard midpoint (which would be 64 miles from either VOR).

This is a perfect example of how the limitations inherent in the VOR system have been accommodated in the Victor Airway System. All the pilot has to do to remain within the limits of the system is to observe the restrictions indicated on the charts.

DOGLEG AIRWAYS

Not all airways are a straight line between VORs. In some cases an airway may change direction, or

dogleg. Course changes are automatic changeover points, and will either be depicted on the chart at an intersection with an obvious change in course (Barry intersection, Fig. 4-1, for instance), or, in the absence of an intersection, with an *x*, indicating a change in course and automatic changeover point (also illustrated in Fig. 4-1, on V-513 between EWO and LVT). The reason the changeover is automatic at the point of the course change is that VOR radials are straight line courses—they can't be "bent;" the only way to make a course change between two VORs is to track outbound on one course and inbound on another.

INTERSECTIONS

Predetermined airway intersections are established for the convenience of ATC and pilots alike in position reporting. Instead of having to describe a fix in terms of radials off two different VORs, or radial and DME off a single VOR, the pilot or controller merely has to use the name of the intersection to describe the position.

Published intersections also provide accurate fixes for the pilot of a non-DME equipped aircraft to use in determining ground speed by dead reckoning. Not only are the published radials accurate to within one degree, but the FAA attempts to use the optimum combination of distance and angle (90 degrees being optimum) in describing intersections for maximum accuracy. In addition, the distances between intersections, or between intersections and succeeding VORs, have already been measured for the pilot to within one nautical mile of accuracy. All of these factors make published intersections excellent fixes to use in determining ground speed.

Intersection Identification—Cross Radial Method

The most common method of identifying Victor Airway intersections is with a cross radial from another VOR. This method of locating present position using intersecting VOR radials was described previously in Chapter 2 in general terms. The principles are exactly the same in Airway navigation, but again, certain restrictions apply.

Intersections are frequently (but not always) formed at the point where two airways cross. Both Renro and Osine intersections (Fig. 4-3) are formed in this manner, for instance. Looking specifically at Osine intersection, notice the two small arrows, one pointing towards the intersection from the north, and the other pointing towards the intersection from the east. These arrows indicate the radials to be used to most accurately identify Osine intersection. In this case, assuming you are already navigating along V-178 either to or from EWO VOR, the 190 degree radial From the Mystic VOR (MYS, 108.2 MHz) should be used as the cross radial for proper intersection identification. (No doubt the 007 degree radial from Bowling Green would also work—the other side of V-49—but less accurately.) If you were navigating along V-49, then the appropriate cross radial would be the 258-degree radial off New Hope VOR.

Mahen intersection is an intersection formed purely for its own sake—no airway crosses V-178 at this point. Instead, Mahen is formed by V-178 and

the 165-degree radial off the Owensboro VOR (108.6 MHz, OWB), which happens to be a low-power, terminal VOR, limited to 25 nautical miles Standard Service Volume, and represented on the charts by a small compass rose and no box around the station information. (As an airway user, you do not need to be concerned with this limitation.) The arrow at the end of the 165-degree radial line from OWB indicates that that is the proper cross radial to use for Mahen intersection, and the arrow on V-178 from Central City VOR indicates that that is the other radial to use. (In this case, since Mahen is so close to Central City, that is fairly obvious, but in other cases it won't be.) In every case, the chart will indicate, by the use of arrows, which radials should be used for optimum intersection identification.

Intersection Identification—DME Method

An optional method of identifying intersections is with the use of DME. Since DME is not required for Airway navigation, cross radials are always provided to identify intersections, but the use of DME, when available, simplifies and adds accuracy to intersection identification.

As an example, look again at Renro intersection (Fig. 4-3). Directly under the arrow on V-178 is a note for "D32." This means Renro intersection can be defined by V-178 and the 32.0 DME fix from Central City VOR.

DME simplifies intersection identification considerably, especially if the DME is auto-tuned to the VOR. With auto-tuning of the DME, dialing in Central City VOR automatically also sets the DME to Central City; when the DME reads 32.0, the aircraft is over Renro intersection. No additional VOR tuning or cross-radial selection is needed. Since DME is more accurate than VOR, this method of intersection identification is more accurate than using intersecting radials, and is the preferred method of intersection identification.

RECEPTION ALTITUDES

The Standard Service Volume for en route low-altitude VORs is 40 nautical miles, which means that the standard maximum useable distance between two VORs is 80 nautical miles. But this is for unrestricted, off-airway use as low as 1,000 feet above the surface. Many airway segments between VORs are more than 80 nautical miles apart; this is done by raising the minimum useable altitudes sufficiently to accommodate the extra distance. The pilot navigating along the Victor Airways does not need to concern himself with the distance between VORs, nor does he need to be concerned with terrain clearance or interference from other VORs with the same frequency, so long as he observes the published altitude restrictions.

Minimum En Route Altitude

Each airway segment has a published *Minimum En Route Altitude* (MEA). The general MEA for the entire segment between VOR fixes is printed directly under the airway identifier—refer again to Fig. 4-3. The MEA for V-178 between Central City and New Hope is 2,700 feet MSL, for instance. This means that the FAA has flight checked this airway segment, and determined that in order to receive VOR azimuth information along the entire route of flight between Central City and New Hope VORs and clear all terrain by at least 1,000 feet, the pilot must cruise at or above 2,700 feet MSL.

Sometimes the FAA determines that certain intermediate segments require a higher MEA than that generally used for the entire segment, and in other cases certain intermediate segments may allow for temporarily lower MEAs. When this happens, rather than require that the entire segment be flown at the highest altitude necessary, the FAA marks on the en route charts, directly under the line for that segment, the MEA for that particular segment.

For instance, Figure 4-5 shows an airway segment, V-16, between Hinch Mountain (HCH, 117.6 MHz) and Knoxville (TYS, 116.4 MHz) VORs, in which one segment has a higher MEA than the rest of the segments: 5,000 feet MSL between HCH and Bucky intersection; 3500 feet MSL between Bucky and TYS. Rather than require that the entire segment between VORs be flown at 5,000 feet, only the affected intermediate segment is restricted to 5,000 or above. This requires careful monitoring of position and altitude by the pilot operating at minimum altitudes, but it does provide him with maximum flexibility in altitude selection. (These are minimum altitudes—the pilot can always voluntarily elect to operate at the higher altitude for the entire segment.)

In other cases, a lower altitude is allowed for an intermediate segment than is allowed for the entire segment. V-4-53, shown in the upper right hand corner of Fig. 4-1 between Louisville (LOU, 114.8 MHz) and Lexington (LEX 112.6 MHz) has an MEA of 2800 feet. (Disregard the + sign before 2,800 on the chart—that is a grid marking; it just happens to

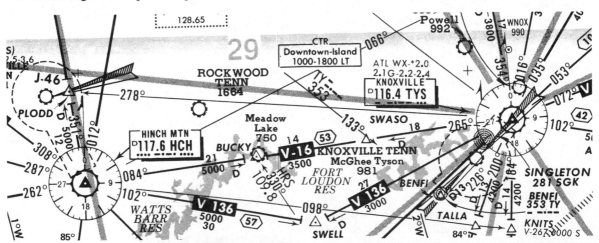

Fig. 4-5. The general MEA for V-16 between HINCH MOUNTAIN and KNOXVILLE is 3500 feet MSL; however, for the segment between HINCH MOUNTAIN and BUCKY intersection, the MEA is 5000 feet MSL. Reproduced with permission of Jeppesen Sanderson, Inc. Not actual size. Not to be used for navigation.

be there and has nothing to do with the MEA.) Notice though, that the intermediate segment between Fedra intersection and the LOU VOR has an MEA of 2,600 feet (as opposed to 2,800 for the rest of this segment). This allows the pilot to cruise 200 feet lower in this segment than he otherwise could, while still maintaining adequate reception and terrain clearance.

IFR altitude clearances are normally issued in thousand-foot intervals, beginning with the next thousand-foot altitude above the MEA; however, clearances as low as the actual MEA can be obtained at the pilot's request, traffic permitting. (Situations where a pilot might want to fly as low as possible while on an instrument flight plan would be to avoid icing conditions, or to fly under turbulent clouds, or to descend below the clouds to make visual contact and avoid having to shoot a full approach.) In those situations where the pilot is operating IFR at minimum altitudes and the MEA increases from one segment to the next, the controller should issue a clearance to the next higher MEA prior to reaching that segment; however, the ultimate responsibility for the observance of MEAs lies with the pilot-in-command, and the pilot should initiate a request for a climb to a higher altitude if the controller fails to issue one.

Minimum Obstruction Clearance Altitude

Minimum En Route Altitudes automatically provide for terrain clearance. In fact, in order to ensure adequate VOR reception, they often mandate a higher altitude than that necessary simply to clear the terrain. When this happens, the FAA also publishes a separate *Minimum Obstruction Clearance Altitude* (MOCA). MOCAs are marked on the Jeppesen Low Altitude En Route Charts with a ''T,'' as can be seen in Fig. 4-1 in the segment between Nashville VOR and Lenon intersection on V-140. The MEA for this segment is 3,000 feet, but the MOCA is 2,400 feet. The MEA is, in other words, 600 feet higher than that necessary just to clear the terrain in this segment.

MOCAs are for emergency use only: they do not provide enough altitude to ensure reception of positive navigational guidance over the entire route segment to which they apply (only within the last 22

nautical miles to the VOR). The MOCA ensures that the pilot will clear the terrain if he stays within the lateral limits of the airway, but without positive navigational guidance the pilot has no way to know for sure that he has remained within those limits—an inherently contradictory situation. This is why MOCAs are for emergency use only. Under normal IFR conditions, MOCAs can and should be disregarded, and the MEA observed instead.

Minimum Reception Altitude

Occasionally, in order to receive a cross radial for a particular intersection, an altitude higher than the MEA is required. In our original example, Fig. 4-1, V-5-49 between Nashville and Bowling Green VORs has an MEA of 2,700 feet; however, a *Minimum Reception Altitude* (MRA) of 3,000 feet applies to the Tands intersection, located in the middle of that airway. (This is indicated by the parenthetical note under the Tands intersection identifier.) Since Tands intersection can also be identified by DME, and since it has little significance except to show a change in the MOCA, rather than require all aircraft to fly the entire airway segment at 3,000 feet just to receive Tands intersection by VOR, the restriction is merely noted; however, if for some reason you want to identify Tands intersection using only the VOR information shown, you must be at or above 3,000 feet to do so.

Maximum Authorized Altitude

On rare occasion, a *Maximum Authorized Altitude* (MAA) will be noted for a particular airway segment directly under the MEA. Since reception distance increases with altitude, the higher the cruising altitude, the greater the number of stations received. In a very few cases the FAA has found that it must restrict the maximum altitude in order to ensure that the desired VOR is not interfered with by another VOR with the same frequency. This is not a common situation when operating on the airways, but you should know what MAA means when you come across it on an en route chart.

JET ROUTES

Jet Routes are a separate system of airways based

on VOR/DME facilities. They are very similar to the Victor Airways, but are reserved for use from 18,000 feet MSL to FL 450. (Since the Flight Levels begin at 18,000 feet MSL in the United States, this means that for all practical purposes the Jet Routes are Flight Level Airways.) Figure 4-6 shows a portion of the Jeppesen U.S. (HI) 4 High Altitude En Route Chart, showing the Jet Routes in the vicinity of Louisville and Nashville.

Jet Route is, in fact, a little bit of a misnomer, since any aircraft capable of operating at FL 180 or above can be issued a clearance along Jet Routes, regardless of the aircraft type. (Actually, if we are going to call airways labeled with a V Victor Airways, to be consistent we should call airways labeled with a J Juliett Airways, but Jet Route seems to have stuck.) In Canada, Jet Routes are called High Level Airways, and are predicated on NDBs as well as VORs. Whatever the name, Jet Routes are open to all types of aircraft.

The line-of-sight reception distance at FL 180 and above is on the order of 200 miles or more, so the distances between VORs tend to be much greater on the Jet Routes than they are on the Victor Airways. Since terrain clearance is not a problem anywhere in the Contiguous U.S. at FL 180 and above, MOCAs for Jet Routes are always something lower than 18,000 feet and are not published. MEAs on Jet Routes are assumed to be FL 180, unless noted otherwise. At these altitudes, terrain-induced interference is not much of a problem, so changeover points are usually either at the midpoint or course change point. Because most aircraft operating at FL 180 and above have DME (and are required to have DME at FL 240 and above), and because all of the airspace at and above FL 180 in the U.S. is under positive radar control, intersection fixes are less common on Jet Routes than they are on Victor Airways.

As a result of these factors, the high altitude en route charts which depict Jet Routes are easier to read

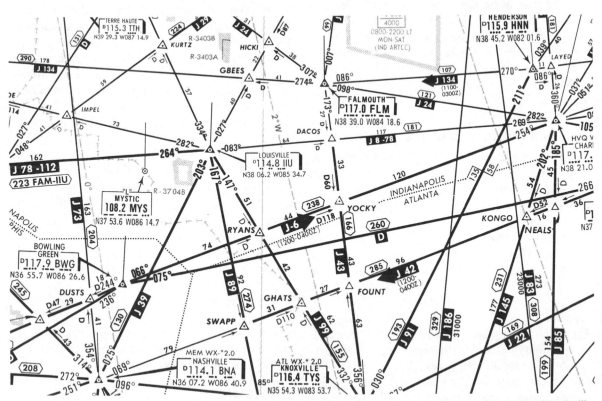

Fig. 4-6. A portion of the Jeppesen US (HI) 4 High Altitude Enroute Chart, showing the Jet Airways in the vicinity of Louisville and Nashville. Reproduced with permission of Jeppesen Sanderson, Inc. Not actual size. Not to be used for navigation.

and understand than the low altitude en route charts, and Jet Routes are simpler and easier to use than Victor Airways. Because of the large distances between VORs, the high altitude charts cover much greater areas on the same piece of paper: two high altitude "Jepp" charts (front and back, for a total of four) cover the entire U.S. Nonetheless, despite their being easier to read and use, there are no essential differences between Victor Airways and Jet Routes, or the charts that represent them—the two are part and parcel of the same VOR/DME based navigational system, and the differences that do exist are simply a function of their differing altitude coverages.

DUAL VOR TECHNIQUES

It is not necessary to have two VOR receivers to navigate on the Victor Airways—intersections can be identified by maintaining heading momentarily, switching to the intersecting VOR, noting the radial, and switching back again. This is a lot of work, but it certainly can be done. Dual VOR receivers are not even required for instrument operations; however, due to the importance of having accurate and continuous navigational information under instrument conditions, most pilots insist on having a back-up VOR receiver.

While the primary reason for having dual VORs is redundancy, as long as the second VOR is there, there is no reason not to take advantage of it. Dual VOR techniques vary from pilot to pilot, and vary depending upon whether DME is available or not. The techniques that follow are just that—techniques. They are not required procedures and pilots are free to develop their own, so long as the results obtained are within system standards.

Without DME

When DME is not available, the primary purpose of the second VOR is position determination. The first VOR is normally used to track the airway, and the second VOR is used to pre-set en route intersections, mark changeover points, and make random cross radial checks as necessary. At the appropriate changeover point, the #1 VOR is usually retuned to the next VOR and the inbound radial set

in the #1 CDI.

Without DME, it is difficult to tell when you are approaching the station, except for the increased needle sensitivity of the CDI (which can be a misleading indication). It is therefore a good idea to set up an intersecting radial 5 to 10 miles in front of the approaching VOR. This intersection will alert you to the approaching station so you will know to get ready for the course change, and it will also reassure you that any erratic needle movements that occur after that point are a function of proximity to the station.

With DME

When DME is available, it isn't necessary to use the second VOR for position determination, since that can be done with the DME. With the second VOR free, you can "lead" your course changes with the #2 VOR. When tracking outbound, the next course (which would normally be inbound) should be set on the #2 VOR; tracking inbound, the course change after the VOR should be set on the #2 VOR. In other words, the #2 VOR always has the next course in it.

For example, to navigate using this technique along V-178 inbound to New Hope on the reciprocal of the 069-degree radial (Fig. 4-3), the #1 VOR would be set to the New Hope 069 inbound (249), and the #2 VOR would be set to the New Hope 258 outbound (V-178 outbound from EWO). Once New Hope had been reached, the #1 CDI would then be turned to the outbound course (258), and the #2 VOR and CDI would be reset to the next VOR and inbound course down the line (Central City, and the reciprocal of 075, or 255).

This system not only enables you to "stay ahead of the airplane" by always having the next course change on the #2 VOR, but it prevents errors in several different ways. First, it provides a double check of the outbound radial at station passage—when you change the #1 CDI to the outbound course, it should match the #2 CDI—if it doesn't, one of them is wrong. Second, if the airway between that VOR and the next one is a straight line course (no dogleg), then as soon as the next VOR station can be identified and the radial received, the course needles of the two CDIs should more or less match—if one is centered and one shows a full scale deflection, one of the radials is probably set wrong. (I say "more or less" because they usually won't match up exactly due to differences in magnetic variation and course width, but they should be close enough to both be on the scale.) Third, if the airway is a dogleg airway, the second VOR will provide a good check on the course change: as the second CDI needle moves towards the center, it indicates that the course change is approaching; when it is centered, you should be at the changeover point. (If it crosses over to the other side of the CDI, you have gone past it.) This system of setting up the next course on the #2 VOR will catch most mistakes in tuning and radial setting, both for straight line airway segments and for doglegs.

RADIO MAGNETIC INDICATORS

The *Radio Magnetic Indicator* (RMI), as its name implies, is an instrument that combines radio and magnetic information to provide continuous heading, bearing, and radial information. (See Fig. 4-7.) It is an extremely simple indicator, and that is both its strength and its weakness: it is easy to obtain information from an RMI, but it is not always easy to know what to do with that information once you have it.

With regular use, the interpretation of radial and bearing information from an RMI becomes second nature—military aircraft are often equipped only with RMIs for NDB and TACAN navigation, and military pilots trained and experienced in their use frequently wonder at the necessity for a CDI at all. Most civilian pilots, however, prefer the CDI for primary airway tracking, using the RMI as a supplement, if at all. (The RMI is not, in fact, a common instrument in the general aviation cockpit except at the heavy twin and turbine level, but it is becoming more so, and with good reason.) An understanding of the way in which it differs from a CDI can be helpful to any pilot.

A note at this point: The "Radio" part of the

Fig. 4-7. An RMI combines radio and magnetic information to provide continuous heading, bearing, and radial information.

Radio Magnetic Indicator can be either VOR or NDB information. Therefore, much of what is said here also pertains to RMI operation for NDB navigation (covered in detail in Chapter 6). The principles of operation are exactly the same, regardless of the type of radio information being displayed; this section will concentrate however, on the use of RMI with VOR information.

Description

An RMI consists of a slaved magnetic heading indicator, and normally two needles, one single barred and one double barred, for the display of radial and bearing information. A *slaved magnetic heading indicator* is a gyro operated compass card that requires no pilot input—a remote sensor detects the magnetic heading of the aircraft at any given moment and relays that information to the RMI where it is displayed, with heading at the top. (It is very similar in this respect to a slaved *Directional Gyro* [DG], and a cross check of the two compass indications is a good check for proper operation of each.)

In most cases, selector knobs are provided so that the pilot may select one of four combinations of VOR and NDB information to be displayed using the two needles: #1 VOR/#2 VOR, #1 VOR/#2 NDB, #1 NDB/#2 VOR, and #1 NDB/#2 NDB. (If only one NDB is installed, then that NDB will be indicated for any NDB selection.) Most pilots—unless the situation dictates otherwise—use the single barred needle to display #1 NDB bearing information, and the double barred needle for #2 VOR radial information. This puts primary NDB information and secondary VOR information—useful for cross radial checks—on the RMI.

Because the RMI always displays current magnetic heading (at the top of the card), the respective pointers (the ends with the arrows) always indicate magnetic bearing To the station—either VOR or NDB. (Magnetic bearing is the same as the magnetic direction or course to the station.) In other words, the pointer not only always points in the direction of the station, but it also indicates the magnetic direction to get there.

If the pointer indicates the magnetic bearing To the station, then the tail must indicate the magnetic bearing From the station, and the magnetic bearing From a VOR is the same as the radial from a VOR.

This is an important point to remember and bears repeating: the tail of an RMI needle always shows the current VOR radial (when set to indicate VOR information, of course). If, for instance, a controller asks you your radial from XYZ VOR, all you have to do is give him the number shown under the tail of the RMI VOR pointer. If you ever get lost or disoriented using an RMI, just remember that your position lies somewhere along the radial indicated by the tail of the pointer.

Principles of RMI Operation —VOR Airways

RMI operation is continuous and automatic: the pointer indicates at any given moment the bearing to the station and the tail indicates at any given moment the radial from the station. This makes the RMI an ideal indicator for cross radial checks, and a good supplement to the DME for position checks. With the RMI set so the needle reflects #2 VOR information, the pilot navigating along a given airway can keep a running tally of his position along that airway by tuning the #2 VOR to an adjacent facility, and watching the RMI needle to monitor his progress along the airway.

Published intersections can be identified in this way by looking up the intersection radial on the chart (344 for Renro, Fig. 4-3, for instance), and waiting until the tail of the needle points to the published radial—344 in this case. With the RMI, there is no need to set the radial into a CDI or worry about orientation. There is also no confusion as to whether you have gone by the intersection yet—if the tail indicates a radial past the published intersection radial, you have gone by it; if the tail indicates a radial before the intersection, you have not.

Tracking with RMI

VOR airways can be tracked with an RMI just as they can with a CDI—in fact, in some ways tracking with an RMI is a little easier than tracking with a CDI. The hard part for most pilots who are not used to tracking with an RMI is figuring out which way to turn to correct back on course (especially outbound), followed by not knowing how much to turn, not knowing when the proper wind correction angle has been established, and not knowing when they are on course.

There are two ways to solve these problems. One is to remember that the tail always shows the present radial from the station. The present radial can then be compared to the desired radial to determine the proper corrective action. (This is much easier to do than it sounds when you have a chart to help in visualization.) The other is to develop a set of rules based on equivalent indications between CDIs and RMIs for any given situation. Both ways work, and most pilots probably find themselves using a combination of each.

Turning directly To a VOR is even easier with an RMI than it is with a CDI. With an RMI, it isn't necessary to turn the OBS until the needle centers with a To indication, since the RMI needle is already pointing to the station. All you have to do is turn the airplane to the magnetic bearing indicated by the needle.

Having done this, and assuming a cross wind of some sort, the airplane will eventually start to drift off course. As it drifts, the aircraft position will change from one radial to another, and this will be reflected on the RMI by a needle swing (just as it would be reflected on the CDI by a needle deflection).

If, for instance, the bearing to the station is 360 degrees, then the aircraft is on the 180-degree radial (the needle will be pointing to 360, and the tail will be on 180). With a wind from the West, the aircraft will drift to the right, and as can be seen from Fig. 4-8, this means that the aircraft will move from the 180-degree radial first to the 179-degree radial, then to the 178-degree radial, and so on. As it does so, the RMI needle will move from 360 at the top and 180 at the bottom to 359 at the top and 179 at the bottom, 358 and 178, and so on. In other words, the pointer will swing to the left. You can see from Fig. 4-8 that the proper direction to correct is also to the left—if you are on the 178-degree radial and want to be on the 180-degree radial, you have to steer left. Turning in the direction of pointer movement is analogous to turning towards the needle deflection on the CDI. Either method—visualization, or turning in the

direction of pointer movement—will accomplish the same result, and that is a correction back to the desired course.

The amount to turn is exactly the same whether you are using an RMI or a CDI—the aircraft doesn't know or care what kind of indicator you are using. If 30 degrees is a good initial correction to bracket a heading using a CDI, then it is also a good correction using an RMI. The RMI needle will have returned to the center position when the tail shows 180 again—the desired radial in this case. (Since there is nothing to "set" on an RMI, you have to remember which radial is the desired radial.) Once the tail indicates you are back on the desired course, you take half the correction out, watch for further drift, apply half of that in the direction of needle movement, and so on, bracketing a heading just as you would with a CDI. When the tail of the needle is steady on the desired radial—180 in this case—that is the same as the CDI needle staying centered, and the proper drift correction angle has been accomplished. The difference at that point between the aircraft heading and the pointer is the drift correction angle.

Station passage with an RMI is obvious: the needle swings around to the tail. To track a specific radial outbound, you first turn in that direction to parallel the course, just as you would when using the CDI. Since the tail always indicates the present radial, comparing present radial to the desired radial will indicate the direction to turn to intercept the outbound course. (See Fig. 4-9.) For those who like rules, paralleling the course and turning 30 degrees or so towards the needle will also put you back on course. You are back on the desired outbound radial when the tail of the needle indicates that you have intercepted the desired radial.

Once on course, you either wait for the tail of the needle to move away from the desired outbound radial, or automatically apply the wind correction from the previous leg and watch for any further needle drift. If the tail does drift off the desired radial, you note the new radial that you find yourself on, visualize where that puts you on the chart, and make the necessary correction back on course. If you don't like to visualize, then use the rule: turn in the direction of needle movement—if the pointer (the end with

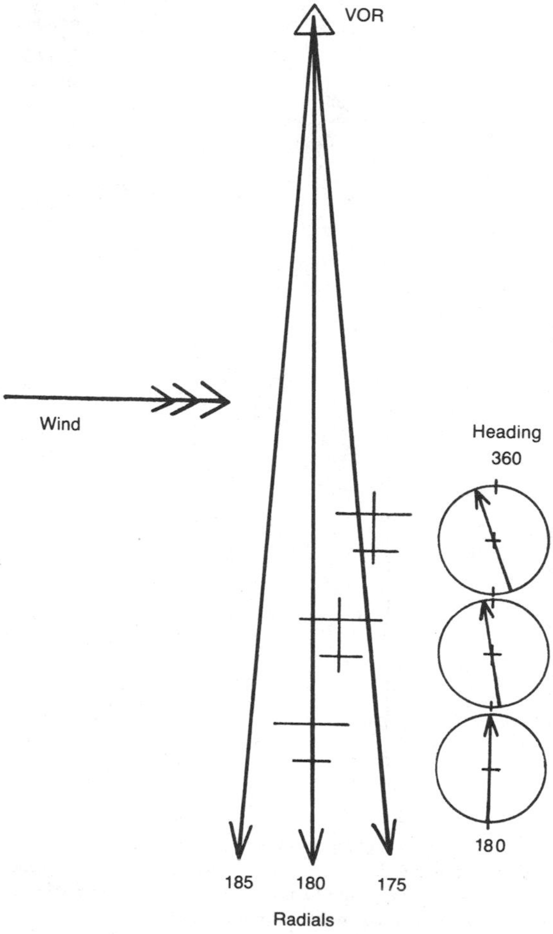

Fig. 4-8. With a wind from the West, an aircraft headed due north on the 180 degree radial will drift over to the 179 degree radial, then the 178 degree radial, and so on. As it does so, the point of the RMI needle will swing to the left.

the arrow) moves to the right, correct right; if it moves left, correct left.

This is the part pilots sometimes have trouble with, because turning towards the pointer when it is at the bottom of the dial seems, intuitively, to be a turn in the wrong direction—it widens the gap between pointer and the bottom of the dial, and this seems wrong. It isn't—the only reason it moves further off is because the heading has been changed. Visualizing your present position, remembering that the tail always indicates present radial, and then making the necessary correction back on course (and remembering that you are back on course when the tail points to the desired radial) is a way of avoiding this problem. Testing the rule in visual conditions often enough that you finally believe it is another. (The other side of this coin is that if you do make a turn to correct, and the tail of the needle gradually shifts even further away from the desired radial, then obviously you turned in the wrong direction.)

To summarize the rules of RMI operation: the aircraft is on-course when the tail points to the desired radial; paralleling the desired course, the proper correction to intercept the desired course is always towards the pointer—the end with the arrow (and if you ever get confused, paralleling the desired course is a good way to get re-oriented). Once on course, if the pointer moves away, turn in the direction of

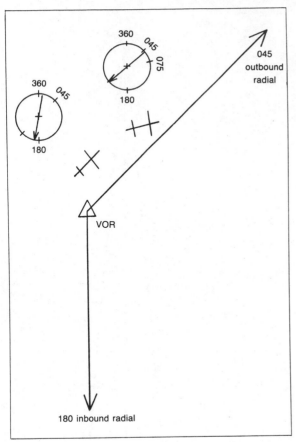

Fig. 4-9. To intercept an outbound radial using an RMI, first parallel the course (045), then make a correction toward the needle. The course has been intercepted when the tail of the RMI needle points to the desired outbound radial—045.

pointer movement. But remember: Once an initial wind correction angle has been established, the direction of pointer movement will be a subtle quality, perhaps moving only very slowly towards or away from the previous position, without crossing over to the other side of the dial. This is why you can't always just turn towards the pointer—you have to turn in the direction of pointer movement—if the pointer is drifting left, for instance, correct left, even if the pointer is on the right hand side of the dial.

Being able to use an RMI for tracking is not only easier in some ways than using the CDI (once you master it), but to a certain extent it is less error prone: there is no chance of confusing the OBS orientation—the pointer is automatically at the top inbound, and at the bottom outbound—and there is no radial to set incorrectly (only one to remember incorrectly). But perhaps the best argument for using an RMI for VOR tracking is that this is the best way to practice NDB tracking—a skill that is rapidly becoming a lost art in many quarters. (See Chapter 6, NDB Navigation, and Chapter 8, Instrument Approaches.)

CONCLUSION

Despite the rapid development of newer systems of air navigation (systems which we will be discussing further on in the book, such as LORAN-C and OMEGA/VLF), VOR/DME based airway navigation will probably continue to be the backbone of the U.S. National Airspace System into the 1990's, and it will almost certainly continue to be used in the less developed parts of the world for many years to come. There is nothing wrong with this though. The VOR/DME system of navigation is easy to use, relatively inexpensive, accurate enough for what it is meant to do, and, with over 30 years of development and experience behind it, most of the problems and limitations have been discovered and resolved. There isn't too much more that can be done with the system, which is why newer forms of navigation are being developed, but there also isn't too much more that needs to be done: day-in and day-out, VOR/DME does the job, and it will be quite some time before the same can be said of any other system.

Chapter 5

VOR/DME
Based RNAV

The VOR system of navigation could be described as a "hub- and-spoke" system: navigational guidance is obtained by moving along spokes, or radials, which extend outward from hubs, or VORs. Because VORs are somewhat randomly and arbitrarily located, seldom arranging themselves in a straight line from departure to destination and therefore forcing deviations from the direct route of flight, the VOR system of navigation is inherently less efficient than a comparable point-to-point system. In the case of the present Victor Airway and Jet Route systems, the average inefficiency is about seven percent—less on longer routes, more on short routes.

The only way to reduce the inefficiency inherent in any hub-and-spoke system is to increase the number of facilities; but seven percent is a fairly reasonable level of inefficiency given the cost and difficulty of reducing it any more with additional VORs, and the gain in efficiency (and therefore the gain in time and fuel saved) in adding VORs has to be balanced against the cost of the facilities themselves. For all practical purposes, the United States has about as many VORs as it needs at this point, and what-

ever inefficiencies result from having to overfly existing VOR facilities will probably continue.

There is, however, a fairly simple and relatively inexpensive way to get around this problem, one that is much less expensive than commissioning more and more VORs (although the question has to be asked, less expensive for whom?), and that is with the addition of a cockpit device called a *Course Line Computer* (CLC) or, more commonly simply an RNAV.

The CLC is a separate cockpit device that takes raw VOR/DME information (i.e., radial and distance information), and turns it into straight-line course guidance between two points (a direct course). In other words, the CLC turns the VOR/DME system, a hub-and-spoke airway system, into an *area navigation* (RNAV) system: a system capable of providing navigational guidance between any two random points within the range and limits of the VORs and DMEs in use.

The acronym "RNAV" (pronounced "R-NAV") by itself is often understood by pilots to mean a VOR/DME-based area navigation system. There are, in fact, many different types of area navigation, or RNAV systems, and in order to avoid confusion among the various types, we will avoid the use of RNAV by itself in this book, except as an abbreviation for area navigation in general.

Exactly how the CLC converts an airway system into an area system is covered in the next section. The bulk of the chapter is devoted to specific ways to put this capability to use, both VFR and IFR. The chapter concludes with a brief review of the limitations of VOR/DME based RNAV, and a look at the prospects for VOR/DME based RNAV in the future.

THEORY OF OPERATION

The theory behind the operation of the CLC is actually quite simple. Any given point that is within range of a VOR/DME facility can be described in terms of a specific radial and distance from that facility. The bearing and distance from any other point (such as the aircraft position) to that point can be determined with simple trigonometry, and the result is a direct course between the two points.

Any point so designated is called a *waypoint*. Waypoints can be described either in terms of radial (theta) and distance (rho), or in terms of latitude and longitude, but, for waypoints based on VOR/DMEs, radial and distance is preferred. The direct course between two waypoints is sometimes called the *Desired Track* (DTK), and its magnetic direction can be found by centering the needle of the associated Course Deviation Indicator (CDI) with a To indication.

With an on-board computer to do the calculations—the CLC—all the pilot has to do is to determine the radial and distance from a convenient VOR/DME to the waypoint desired and enter this data into the CLC. He will then be able to proceed directly from his present position to the desired position—the designated waypoint—bypassing the VOR itself. The CLC keeps a running tally of the aircraft position in relation to the waypoint and the DTK, and indicates the need for any corrections caused by drift or heading error through the CDI.

Most VOR/DME-based RNAV systems use the #1 VOR CDI to show deviations from the DTK—the same CDI that is used for normal VOR tracking. Because the same CDI is used, it often appears, to

the pilot, that the VOR has been "moved" to the waypoint location, and for this reason the waypoint is sometimes called a phantom VOR. What the CLC is actually doing is keeping tabs on the aircraft, comparing the aircraft's actual VOR/DME position with the calculated VOR/DME position between waypoints. This may seem like an academic distinction, but it may help later in understanding why tracking To and From a waypoint is not quite the same as tracking To and From a VOR.

METHOD OF OPERATION

There are any number of different models of VOR/DME-based RNAV systems, ranging from basic, single waypoint units to more complex systems capable of storing entire RNAV routes with as many as 30 waypoints. Some units have self-contained VOR and DME units and can operate independently from the normal navigational systems. Many have extra features such as ground speed calculation, time-to-waypoint, and digital bearing, and some are even able to utilize TACAN information for additional coverage.

Despite the many differences in specific capabilities among VOR/DME-based RNAV systems, all units have essentially the same basic method of operation: the CLC determines present aircraft position internally, while the pilot manually enters the radial, distance, and, in some cases, the frequency of the associated VOR/DME. This means that the aircraft must be equipped with both a VOR and a DME, and those units must be capable of supplying radial and distance information in the proper format to the RNAV unit. In addition, there has to be some mechanism for entering radial and distance information into the CLC.

Figure 5-1 shows an example of one of the more advanced types of VOR/DME-based RNAV systems. This unit has a self-contained VOR receiver, eight waypoint capability, and provides digital bearing and distance to the waypoint. The radial from the selected VOR to the waypoint is shown in the block marked "RAD" (263.3). The distance from the VOR to the waypoint, 087.3 nautical miles (all RNAV systems work with nautical miles), has also been entered, and is shown over the block marked "DST," for distance.

(It is very difficult to measure radials to within a tenth of a degree of accuracy or measure distances to within a tenth of a nautical mile, however, numerous commercial publications list waypoints for airports and other common fixes that have been calculated to within a tenth of a unit, and most CLCs accommodate this level of accuracy.)

Methods of data entry vary with the type of unit. With this particular unit, data is entered with the large knob on the right, after moving the arrow (with the key marked "Data") to the block to change. (In Fig. 5-1, the arrow is shown in the frequency block. Turning the large knob will change the VOR frequency in this case.) Some RNAV units use separate knobs for each parameter, while others use a keyboard for data entry. For the unit shown in Fig. 5-1, once the data has been entered and verified as correct, the button marked "RNV" is pushed, and bearing can then be determined either by rotating the OBS knob on the related CDI until the needle centers, or by pushing the button labeled "Push WPT BRG/DST" to show digital bearing and distance to the waypoint.

There is one important difference between the way deviation from the DTK is indicated on the VOR CDI, and the way deviation from a selected VOR radial is displayed. With conventional VOR navigation, the closer the aircraft is to the station, the more sensitive the needle gets, but with RNAV, the amount of needle deflection does not vary with distance to the waypoint, but always represents a constant distance off track for a given deflection. If a one dot deflection equals one mile of deviation from the DTK (not uncommon), then one dot will always equal one mile deviation, regardless of distance from the waypoint. The reason for this is that a waypoint is not a VOR, however much it might look like one to the pilot. A waypoint is entirely a creature of the computer's imagination, and the computer has the capability to make the course width, and therefore the needle deflection, constant at all distances from the waypoint.

With a constant course width, the pilot can correlate specific needle deflections with specific distances off track. Another advantage to VOR/DME RNAV is that RNAV courses are computed courses, a step removed from raw VOR/DME data. RNAV courses are therefore steadier than VOR courses, and this makes for easier tracking and better rides, particularly when the autopilot is coupled to the nav system.

The next two sections examine various practical applications for VOR/DME based area navigation, including navigation along random routes, and techniques for supplementing other forms of navigation.

VFR APPLICATIONS

Probably the most common VFR application for VOR/DME based RNAV is as an airport finder. Many airports regularly used by VFR pilots are small and can be difficult to see from any distance—they tend to blend in with the surrounding terrain, and the runways, to the extent they can be seen at all, look like roads. By establishing a waypoint over the air-

Fig. 5-1. A sophisticated VOR/DME based RNAV unit with eight waypoint capability, internal VOR receiver, and digital waypoint bearing and distance readout. Photo courtesy of Narco Avionics.

port, the pilot can, in effect, put a VOR on the field, which is a great help in locating and positively identifying airports in unfamiliar areas.

To create a waypoint, the pilot should first draw a line on his Sectional chart from a nearby VOR to the airport. The radial from the VOR to the waypoint (the airport) can then be read directly off the compass rose around the VOR—there is no need to use a protractor or convert True Bearings to Magnetic, because the VOR compass rose is already oriented toward Magnetic North. Distance, in nautical miles, from VOR to waypoint is then measured on the chart with a plotter. With radial and distance from the VOR to the airport available, the pilot has established the two essential elements necessary to establish a waypoint.

As soon as the aircraft is within reception range of the selected VOR, the radial and distance for the waypoint are entered into the CLC and the RNAV function selected. The CDI will then provide course guidance directly to the waypoint—the airport in this case—negating any need to fly first to the VOR and track outbound to the airport, or rely on dead reckoning to locate the airport.

This airport finding capability is one of the most common, and most useful, applications for VOR/DME-based RNAV systems for the VFR pilot, and it can be accomplished with basic, single waypoint RNAV systems.

Random Routings

VOR/DME-based RNAV can also be used by VFR pilots to provide direct Rhumb Line courses between any two points—random routings—just as is done with dead reckoning. This is a little more complicated than the previous application, and it works best with an RNAV unit with more than one waypoint capability; it is not, however, difficult to do, it can be done with a basic unit, and it can provide all the time- and fuel-saving efficiency of direct courses based on dead reckoning without the uncertainties inherent in those procedures.

To plot an RNAV route between two points, the pilot first plots a direct course just as if he were navigating via dead reckoning. He then locates several VOR/DMEs along the route of flight, and

draws intersecting lines from those facilities to the direct course in order to create a series of waypoints along the route. (The only VORs that cannot be used for VOR/DME-based RNAV are those that have neither DME nor co-located TACAN. See Chapter 4, VOR/DME Applications, for a discussion of the differences among VOR types.) For convenience, cardinal directions, such as 360 or 090, are commonly used for the waypoint radials, but other radials can be used as well. The distance from each VOR along the radial selected to the course line is then measured, and the preflight navigational planning is complete. En route, the pilot enters the waypoints into the CLC, in order, and navigates directly To the waypoints, creating a straight line course where none existed before.

A VOR/DME-based RNAV system with advanced features can make this process even easier. For instance, many units allow the pilot to navigate From a waypoint, as well as To. Thus, over the first waypoint, instead of having to immediately jump ahead to the next waypoint, the pilot could navigate outbound from the original waypoint. If the RNAV unit has the capability to store waypoints, then each of the waypoints for the entire route of flight (plus a couple of alternates) can be entered and stored ahead of time for recall and use in the air. Some units even have back-up batteries, enabling frequently flown routes to be permanently stored.

RNAV systems can be very helpful in VFR operations, both in locating difficult to find airports, and in reducing airway distances with direct routings. To the extent direct routings increase efficiency, they also reduce expenses. In practice the savings can be elusive, however, and the greatest benefit of VOR/DME-based RNAV for the VFR pilot is probably its ability to provide positive and reliable navigational guidance directly to specific fixes, such as airports—any cost savings are an added benefit.

IFR APPLICATIONS

VOR/DME-based area navigation for IFR has never achieved the wide-spread popularity that it was assumed it would attain when first introduced. This is not because VOR/DME-based area navigation has not been up to the job; rather it has been mainly due to the inability of the instrument system to accommodate it.

The first priority of Air Traffic Control (ATC) is aircraft separation, not navigation. ATC has a very difficult time with separation when aircraft operate along random routes, especially in terms of coordination from one sector or center to another. (Perhaps this will change when the next generation of ATC computers comes on line, but it will be several years before that system is operational.)

In fact, the only serious attempt by the FAA at integrating VOR/DME-based area navigation into the IFR system was a network of area routes called the U.S. RNAV Route System. An RNAV route system is a little bit of an oxymoron—How can an area system have routes?—but such a thing did exist in the form of direct courses between certain major city pairs with predetermined waypoints. With the exception of a couple of RNAV routes in Alaska, the U.S. RNAV Route System no longer exists. It failed for fairly predictable and obvious reasons: it was too limited to be very useful, and it was too difficult to expand without running into problems with aircraft control. The ATC system, as it presently exists, needs airways to manage traffic.

This is not to say, however, that approval for random, direct IFR routings based on VOR/DME facilities is not possible: IFR area navigation in the National Airspace System is possible, but, in practice, it is a rarity.

This section will look first at what is required to obtain an IFR clearance for a random route, should you have the curiosity or need to try one, followed by a look at some of the many ways VOR/DME based RNAV can be used to effectively supplement conventional VOR/DME navigation—by far its most common IFR application.

Random Routings—IFR

The *Airman's Information Manual* (AIM) states, under Area Navigation (RNAV) Routes (paragraph 298 d. [2]):

> The complexities involved in determining route width, with reference to facility usable distance, requires that random RNAV routes only be approved in a radar environment. ATC will radar monitor each flight, however, navigation on the random RNAV route is the responsibility of the pilot. Factors that will be considered by ATC in approving random RNAV routes, include the capability to provide radar monitoring and compatibility with traffic volume and flow.

Thus, the first requirement to obtain a random route IFR clearance is radar coverage. Most of the United States is covered by radar above about 5,000 feet AGL, so this is not a seriously limiting factor, but it may force a higher cruising altitude than would otherwise be indicated. The last phrase in the paragraph regarding "compatibility with traffic volume and flow" is by far the most seriously limiting factor, and is the reason IFR RNAV is not common.

The *AIM* also lists (Para 298 d. [3]-[5]) the specific requirements necessary to obtain an IFR RNAV clearance. They are, somewhat condensed and paraphrased:

1. File IFR airport to airport, using the appropriate suffix for RNAV capability (i.e., slant Charlie, Romeo, or Whiskey—see *AIM* para 298 a. [3] for a description of all suffixes).

2. The route must include standard departure and arrival routings (SIDs, STARs, and other "appropriate arrival and departure transition fixes"), and the random route portion must fall between those departure and arrival points.

3. File random routes in terms of degree/distance waypoints, at least one per center, and at least one waypoint in each center area must be no further than 200 nautical miles from the edge of the previous center.

4. Plan additional waypoints for any turn points and to insure navigational accuracy (which remains the responsibility of the pilot).

5. The route must circumnavigate Prohibited and Restricted Airspace.

6. The RNAV equipment being utilized must meet the requirements of AC 90-45, an Advisory Circular issued by the FAA, having the force of regulation. (AC 90-45 essentially describes the requirements to obtain IFR approval for RNAV equipment.)

It is fairly easy to see, given the long odds of obtaining an IFR random route clearance in the first place, and considering the requirements above for filing an RNAV IFR flight plan, why few pilots bother to do so. It can be done though, and my guess is that with a little experimentation and practice it might not be so difficult to file an RNAV flight plan or obtain an IFR RNAV clearance as it seems.

Supplementary Applications

The real question, from a practical point of view is, what can VOR/DME-based RNAV do for the instrument rated pilot to make his job of flying from point A to B easier and safer? There is much that it can do, as explained in the following sections. A general caveat regarding IFR RNAV applies, however; in each of the cases that follows, there is a legal requirement that the RNAV unit meet the specifications of AC 90-45. (*AIM*, para 298 d. [5]: "To be certified for use in the National Airspace System, RNAV equipment must meet the specifications outlined in AC 90-45.") You can always turn a non-approved RNAV unit on in instrument conditions and monitor its performance, but you are not allowed to rely on its information for IFR navigation, nor file as RNAV capable, unless it is IFR approved.

Direct Routings. Ironically, one of the most common uses in the air for VOR/DME-based RNAV, is for exactly that which seems so difficult to obtain on the ground: a direct clearance to a point well ahead of the aircraft. The real problem ATC has with RNAV clearances is in granting the clearance ahead of time, while still on the ground; once in the air, dealing with the situation that actually exists at the time, all sorts of things are possible that are not possible on the ground, including the granting of long, direct routings.

When ATC sees one of the RNAV suffixes attached to the aircraft identifier on the flight plan (normally "/R"), it assumes long-range area navigational capability and may issue a clearance to a point many hundreds of miles ahead of the aircraft, just as if the aircraft were equipped with LORAN-C, or OMEGA, or INS (long-range systems covered in later chapters). When this happens, the quick thinking pilot can use his VOR/DME-based RNAV capability to draw a course line from his present position to the distant fix, plot waypoints along the way, and in this way proceed as cleared.

Since it usually takes a minute or two to draw and plot a series of waypoints and then enter them into the RNAV unit, it is often a good idea to ask ATC for an initial heading to the fix while you plot waypoints and program the RNAV unit. This is perfectly acceptable and well within the capabilities of the controller. (See Chapter 7, Ground Based Radar Navigation, for a review of the use of radar vectors in long-range navigation.) Once the RNAV unit has been programmed, inform the controller that you are proceeding directly to the fix cleared.

Navigation to Intersections. Another use for RNAV under IFR is to facilitate navigation to intersections. It is not at all uncommon, especially in major metropolitan areas, to receive a clearance to proceed directly to an intersection. Without RNAV capability, this can be a difficult maneuver, since an intersection is not a facility that can be tuned, but is simply a place defined by other facilities. To find a VOR intersection without RNAV equipment, it is necessary to first intercept one of the radials defining the intersection, and then follow that radial to the intersection itself. An intersection is normally one of those places "You can't get to from here"—you have to get onto one of the radials forming the intersection so you can "get there from there."

With RNAV capability, all the pilot has to do is enter the radial and distance from one of the VOR/DMEs defining the intersection into the RNAV unit, select RNAV, center the CDI, and proceed directly to the intersection. With the needle centered, the aircraft will be over the intersection when the DME reads zero. The pilot can then re-select VOR and continue to navigate normally according to the rest of the clearance. Navigating directly to an intersection using RNAV reduces pilot workload, and expedites the arrival.

Visual Approach Assistance. In a similar fashion, RNAV capability can be put to good use when issued a clearance for "vectors for the visual:" assigned headings in the expectation that the aircraft will be brought close enough to the airport to spot it visually, (or, if there is another aircraft ahead which has already been given a visual clearance, close enough to spot and follow that aircraft). If the airport does not have a VOR/DME on the field, but does have a VOR/DME located nearby, then a waypoint can be established over the airport and entered into the RNAV. The RNAV will then show distance and bearing to the airport—a great help in finding and positively identifying the airport when other navigational aids on the field are not available.

Distance to Touchdown Indicator. Likewise, on an instrument approach (see Chapter 8 for a complete discussion of instrument approaches), the RNAV unit can be set up to provide distance and bearing information to the airport. (This assumes that the RNAV system is independent of, or at least does not conflict with, the VOR/LOC receiver being used for the approach.) For instance, with the #1 NAV set for an ILS approach, the RNAV can, in some cases, be selected to accept radial and distance information through the #2 VOR (or, from its own independent VOR receiver and DME). With the airport programmed as the waypoint, and RNAV selected, the DME will then show distance to the airport, and bearing to the airport will be shown on the #2 CDI. This is valuable information, both while on vectors for the approach, and on the approach itself.

It is worth emphasizing though, that RNAV information cannot be used for the approach itself (unless an RNAV approach is specifically depicted and approved), nor can RNAV information be substituted for required approach components such as compass locators or markers. It can only be used as a supplement to assist in planning and orientation.

NDB Airways. VOR/DME-based RNAV can be very helpful when navigating along an NDB airway. (NDB airways are commonly found only outside of the United States. See Chapter 6, NDB Navigation.) Most pilots have trouble navigating along NDB airways, mainly because they have little experience or practice in tracking with an ADF card or RMI. One way to get around this is to substitute an RNAV waypoint for each of the NDBs that make up the NDB airway, and then navigate from waypoint to waypoint, while monitoring the ADF indications. Legally, the primary navigational guidance still has to come from the NDB when cleared along an NDB Airway, but

with the RNAV available to show the way and indicate directly, positively, and in a familiar manner any deviation off course, NDB tracking is made much easier, and many awkward mistakes are prevented.

Airway Cross-Checks. Another airway application for RNAV is to provide accurate en route cross-checks. This is particularly helpful along lengthy en route segments where navigational accuracy in the middle sections may be marginal. J-86, for instance, crosses the Gulf of Mexico from Sarasota (SRQ, 115.2 MHz) to Leeville (LEV, 113.5 MHz), a distance of 414 nautical miles. (See Fig. 5-2.) As it works its way across the Gulf, it passes between several warning areas used by high speed military aircraft for training. This would not be a good area to wander off the airway. By creating waypoints along the route of flight, in particular by creating waypoints for each of the en route intersections (Covia, Nepta, and Santi), an accurate cross-check of airway tracking can be maintained. (The furthest VOR/DME used to define these intersections, Tallahassee [TLH, 117.5 MHz] is only 174 nautical miles away at the airway midpoint, and the VOR/DMEs used for the other waypoints are closer yet.) Primary navigation still has to come from the VORs that define the airway, but this is an excellent back-up technique.

VOR/DME-based RNAV is, in fact, very helpful in any IFR situation where additional checks on position and tracking are desirable. The situations where RNAV guidance can be applied are literally countless. The irony is, that while VOR/DME based RNAV has never been able to realize its true potential in IFR operations in terms of area navigation, it has done more than was ever expected or planned for it in terms of supplemental course guidance in normal airway operations.

LIMITATIONS

The limitations of VOR/DME-based RNAV are essentially the same as those for VOR/DME navigation itself. (VOR/DME-based RNAV is not a new system—it is an add-on to an existing system.) This means that VOR/DME-based waypoints can be created with certainty only when they fall within the Standard Service Volumes of the VORs and DMEs

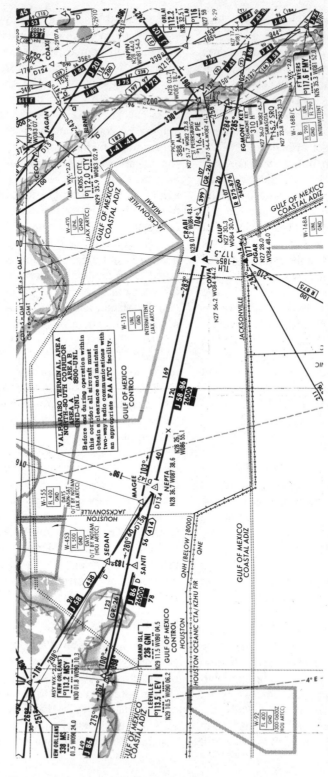

Fig. 5-2. RNAV capability can be used for supplemental course guidance, and is particularly helpful over long enroute segments with critical areas to either side, such as J-86. Reproduced with permission of Jeppesen Sanderson, Inc. Not actual size. Not to be used for navigation.

upon which they are based: normally 40 to 130 nautical miles, depending upon the class of VOR and the aircraft altitude. (For a complete discussion of the limitations of VOR and DME, refer back to Chapters 2 and 3.) RNAV waypoints can sometimes be designated at greater distances than this, but the practical limit is somewhere around 190 to 200 nautical miles from the associated VOR/DME at the highest altitudes normally used in civil aviation. Consequently, VOR/DME-based RNAV is a fairly short-range type of area navigational system, although, as we have seen, it can be used for long-range navigation (over land areas) by joining waypoints together to form longer routes.

OUTLOOK FOR VOR/DME-BASED RNAV

Air navigation is changing rapidly at the time of this writing, and it is difficult to predict exactly what the future will bring. Certain trends do seem to be indicated though, and one trend seems to be a shift away from further development of VOR/DME-based RNAV, toward development and expansion of other, more capable, systems of area navigation, especially LORAN-C and NAVSTAR/GPS (see Chapters 9 and 12). There are several reasons for this shift, not the

least of which is cost: some of the new LORAN-C units now appearing on the market cost less than basic CLCs for VOR/DME RNAV; NAVSTAR/GPS receivers are expected to cost about the same as a high quality VOR/DME unit. Unless the manufacturers can find a way to produce RNAV computers at a very small incremental cost to the VOR/DME units themselves, it is likely that VOR/DME-based RNAV will go the way of the radial engine.

This is not to say that VOR/DME-based RNAV is obsolete at the present time, or that the many pilots who have invested in such units have wasted their money—the uses we have described in this chapter for VOR/DME-based RNAV will be around as long as VORs and DMEs are around, and the VOR/DME system of navigation and air traffic control will be around until that time—several years away as a minimum—when inexpensive long-range navigational systems cover the entire United States and are accepted for normal en route navigation. It is pretty clear though, that VOR/DME-based RNAV has reached its limits in terms of sophistication and growth, and its time as a primary system of area navigation for general aviation has probably come and gone.

Chapter 6

NDB Navigation

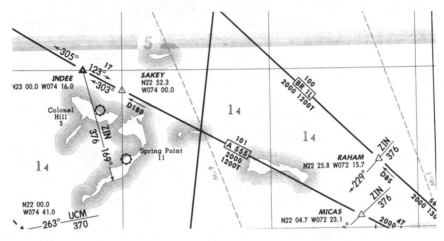

NDB navigation—navigation using Non-Directional Beacons—is the oldest form of electronic navigation still in regular use. NDB navigation is not without its shortcomings or difficulties, and navigation via Non-Directional Beacon is much less common than it used to be, but the virtues of NDB navigation are such that it will probably never completely disappear.

An NDB is a very basic facility. In fact, an NDB is little more than an AM radio transmitter. The reason for its continued use as a navigational aid, when many other systems (such as VOR and LORAN) offer so many advantages over the relatively primitive NDB is simple: NDB facilities are inexpensive to build and maintain, and the airborne equipment associated with NDB reception is only a little bit more complicated than a pocket radio—an AM receiver, a couple of antennas, some circuitry to convert signal strengths to bearing information, and a simple indicator.

En route navigation via NDB is virtually unknown in the United States (except for Alaska), even for VFR navigation, but NDB en route navigation is still common in many other parts of the world, and

it is still a very useful means of over-water navigation at intermediate distances—200 to 600 nautical miles. NDB also provides the only means of navigational approach information for many smaller airports, and NDB assistance, in the form of a Compass Locator beacon, is an important component of most ILS approach facilities. (See Chapter 8, Instrument Approaches, for information on NDB approaches.) For these reasons, pilots should be familiar with NDB navigation, even as newer and more sophisticated forms of navigation push NDB navigation further into a supporting role.

PRINCIPLES OF OPERATION

Any AM-radio facility that transmits a signal in all directions is a potential Non-Directional Beacon, including AM-broadcast stations. Those reserved for air navigation operate primarily in the Low and Medium Frequency (L/MF) Band from 190 to 535 kHz—just below the start of the AM-broadcast dial. NDB transmitters vary in power from less than 25 watts to 2,000 watts, and range varies from less than 15 nautical miles to several hundred, depending on transmitter power and atmospheric conditions. (The nominal range of the most powerful transmitters, called HH facilities, is 75 nautical miles, but they can frequently be received at much greater distances.) The power of the transmitter is always proportional to its intended use, and is not something the pilot needs to be concerned about.

The airborne unit used to navigate via NDB facilities is called an Automatic Direction Finder (ADF) and consists of a receiver, a sense antenna, a loop antenna, and an indicator. (See Fig. 6-1.) It is an *automatic* direction finder because the determination of direction to the transmitting facility is automatic and does not require any antenna rotation or interpretation of relative signal strengths on the part of the pilot: the loop, or directional antenna, is rotated electronically, rather than mechanically, and the bearing to the station is internally derived by combining information from the loop and sense antennas. The result of these internal computations is direction, or bearing, to the station.

Bearing to the station is displayed on an indicator consisting of a compass card and a pointer. There are three types of compass cards, with varying degrees of sophistication and cost: Fixed Card, Rotatable Card, and Radio Magnetic Indicator (RMI). If the compass card is fixed, the bearing displayed is called *Relative Bearing*, because the bearing is relative to the longitudinal axis of the aircraft (which means it is also relative to the heading of the aircraft). If the compass card can be rotated to coincide with aircraft heading, then the bearing displayed is *Magnetic Bearing:* Magnetic Bearing To the station under the pointer, and Magnetic Bearing From the station under the tail. The RMI (previously discussed in Chapter 4) is simply a rotatable indicator in which the compass card is automatically rotated, thus displaying continuous Magnetic Bearing information without the need for any pilot input. Rotatable card indicators are illustrated in Figs. 6-1 and 6-2. A fixed-card indicator always has 0 at the top and 180 at the bottom, and lacks a knob for rotating the compass card, but is otherwise similar. An RMI was illustrated in Chapter 4, Fig. 4-7.

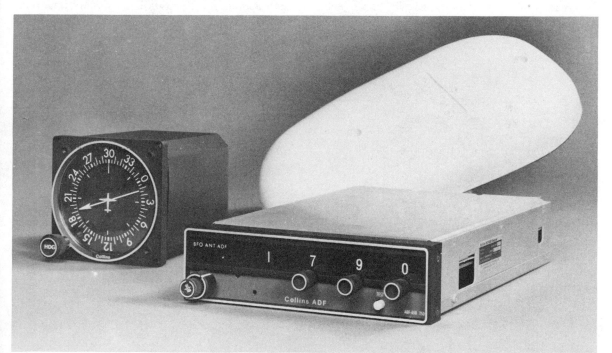

Fig. 6-1. A typical NDB receiving and indicating system: on the left, a rotatable card indicator, and on the right, a panel mounted ADF receiver; to the rear, combined sense and loop antennas in a flush mounted housing. Photo courtesy of Collins Avionics, Rockwell International.

ADF OPERATION

ADF operation is fairly simple (unlike Manual Direction Finding, which is, or was, an enormously complex process). A selector switch, typically marked Off/ADF/ANT/BFO, is required to optimize tuning. (See Fig. 6-2.) The ANT (for antenna) position provides the maximum receiver sensitivity and selectivity and should be used to tune the receiver to the desired station. Newer ADF receivers are digitally tuned and, after selecting ANT, require only that the proper frequency be entered to receive a station—the ADF in Fig. 6-2, for instance, is tuned to 1180 kHz. Tuning older, analog units is a little more complicated, since these units typically have three frequency bands, and the proper band must first be selected. Once the appropriate band has been selected for the frequency desired, the tuning dial must be rotated to the area on the dial of the desired frequency and the signal strength meter (or tuning meter) used to locate the signal. The station must then be fine-tuned for maximum signal strength, as indicated on the tuning meter.

After tuning the station, and with the selector knob still set to ANT, the volume control should be turned up and the station positively identified by listening to the coded ident signal. Codes for beacons suitable for air navigation are provided on the charts. In the U.S., NDB idents generally consist of a three-letter code similar to a VOR ident, except for Locator Beacons, which have a two-letter identifier; in other parts of the world, NDB idents can be either a two- or a three-letter code. In some parts of the world it may be necessary to switch to the *BFO* position (Beat Frequency Oscillator) to hear the ident. The BFO circuit adds a tone to an unmodulated signal, making it easier to identify. (Almost all beacons in North America are modulated.) The BFO position is sometimes also labelled CW, for Continuous Wave, which is another way of saying unmodulated signal.

It is very important that the NDB facility be positively identified. Frequency selectors, both digital and analog, can shift in calibration, and the frequency of ground facilities can also shift as a result of misalignment and atmospheric conditions. Simply tuning the receiver to the proper frequency is not a guarantee that the proper station has been selected; positive identification of the station is even more important in NDB navigation than it is in VOR navigation. Navigating to the wrong NDB is a common error,

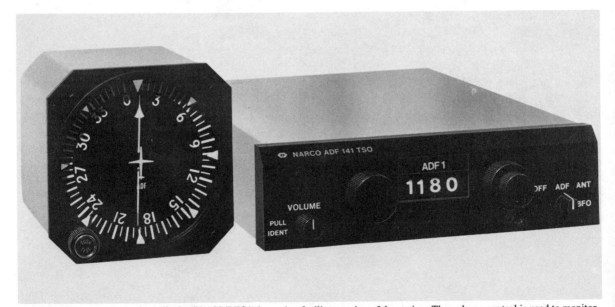

Fig. 6-2. The switch marked "OFF/ADF/ANT/BFO" is used to facilitate tuning of the station. The volume control is used to monitor the ident signal in order to verify proper tuning and functioning of the NDB. The heading selector knob for the compass card includes a "Push-to-Test" function which swings the needle off its last setting. Photo courtesy of Narco Avionics.

even with digital receivers.

Once the station has been properly tuned and identified, the selector switch should be switched from ANT to ADF—Automatic Direction Finding. This is very important. Bearing information will not be displayed unless the selector switch is in the ADF position.

Since there is no On or Off flag on an ADF receiver or indicator, detecting signal interruption on the basis of needle indications alone is nearly impossible: if the signal fails, the needle will usually remain in its last position, and it may be some time before the pilot realizes something is wrong simply because the needle has not moved. (Some units ''park'' the needle in the horizontal position if the signal fails or drops below a certain level, but this is hardly a positive indication of signal interruption either.) The only way to be sure that you are receiving a valid signal is to continuously monitor the ident anytime an NDB is being used. The ident volume should not be turned up so high that it interferes with communications, but it should be loud enough to be noticeable. If the ident disappears, you can try switching to ANT again, and attempt to retune the station. If these efforts fail, the station must be abandoned, even if the information appears to be valid.

Most ADFs also incorporate a Test function. Normally, all the test function does is swing the needle. If the needle doesn't swing when tested, the unit is not working properly. If it does swing but doesn't return to its previous position, then the signal is too weak to be usable. The Test position should always be momentarily selected after the station has been identified and ADF has been selected to verify that a usable signal is being received. It can be re-selected anytime to check the validity and quality of the signal—the faster the needle returns to its previous position, the better the signal.

NDB TYPES AND LOCATIONS

NDBs fall into four major categories and are found at four different kinds of locations. The first, and by far the most common type (in the U.S.), is the Compass Locator, found at either the outer or middle marker on an ILS approach (to be covered in Chapter 8). These are very low powered facilities, intended solely as approach orientation aids.

The second type of NDB is the NDB approach facility, found at or near airports where NDB is the primary approach aid. These beacons are slightly more powerful than Compass Locators, with a standard range of 25 nautical miles, but are still not intended to be used for en route navigation. NDB approaches are also covered in Chapter 8.

The third type of NDB is the en route airway beacon. These are medium powered transmitters with a standard range of 50 nautical miles. En route airway beacons are quite common in other parts of the world, including Canada and the Caribbean. NDB airways are called L/MF Airways (for Low/Medium Frequency), and are identified by color and number. Amber and blue airways run generally north and south, and red and green airways run east and west. A typical L/MF Airway is illustrated in Fig. 6-3.

The last type of NDB is the high-power beacon. High-power beacons are generally located along the coast; their purpose is to provide navigational assistance between the shore and various oceanic fixes, such as entry points to over-water routes, and between the shore and island-based facilities. Figure 6-4 shows an example of a high-power NDB airway, A-22 (Amber 22), from the Nantucket beacon (TUK, 194 kHz) to Slatn intersection. Slatn is an initial transoceanic fix commonly used as a jumping off point for Bermuda.

NDBs are seldom used on long-range oceanic routes anymore (except occasionally by ferry pilots or in emergencies), having been replaced by much more effective long-range systems, however, NDBs are still extensively used offshore for island navigation throughout the Bahamas and Caribbean. Figure 6-5 shows several such routes off the Carolina Beach and Dixon beacons (216 kHz and 198 kHz respectively). These are called Atlantic Routes (AR-1, AR-3, etc.), and they connect the U.S. both with other points in the U.S. (AR-1 terminates in Florida) and with Bimini and Nassau beacons in the Bahamas. In some cases these long range beacons are co-located with a VOR for greater accuracy close to the end of an airway, but the primary en route navigational aid over these longer distances (AR-3 is 545 nautical miles long) is NDB.

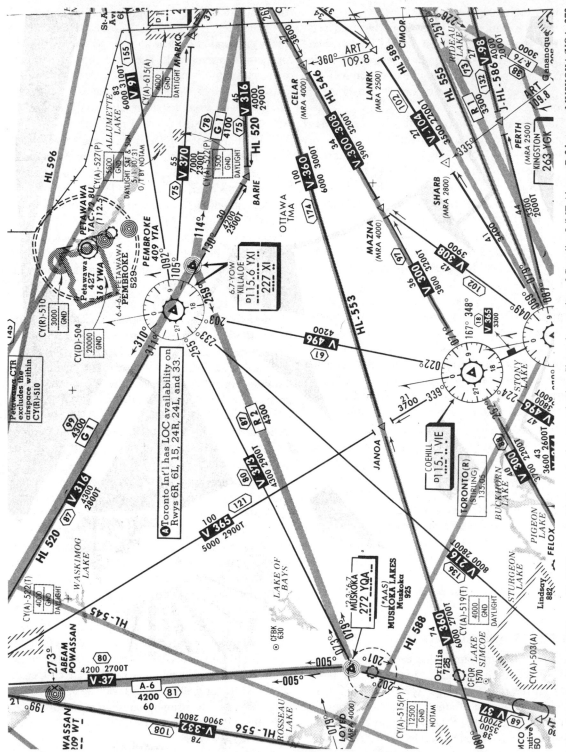

Fig. 6-3. In many parts of the world, NDBs are used for enroute airway navigation. Shown is a section of Canadian L/MF Airway R-2 between MUSKOKA(YQA, 272 kHz) and KILLALOE (XI, 227 kHz) beacons. Reproduced with permission of Jeppesen Sanderson, Inc. Not actual size. Not to be used for navigation.

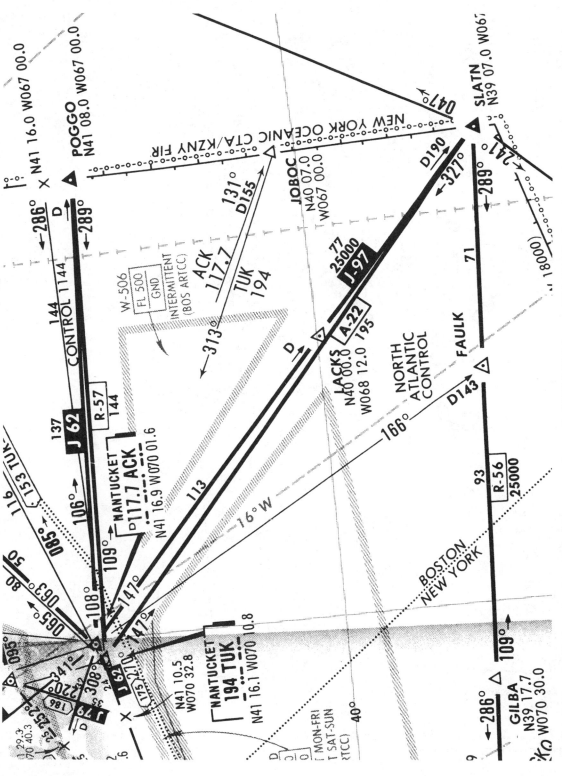

Fig. 6-4. A-22 is an L/MF Airway based on a high power, overwater beacon (the NANTUCKET beacon, TUK, 194 kHz). Reproduced with permission of Jeppesen Sanderson, Inc. Not actual size. Not to be used for navigation.

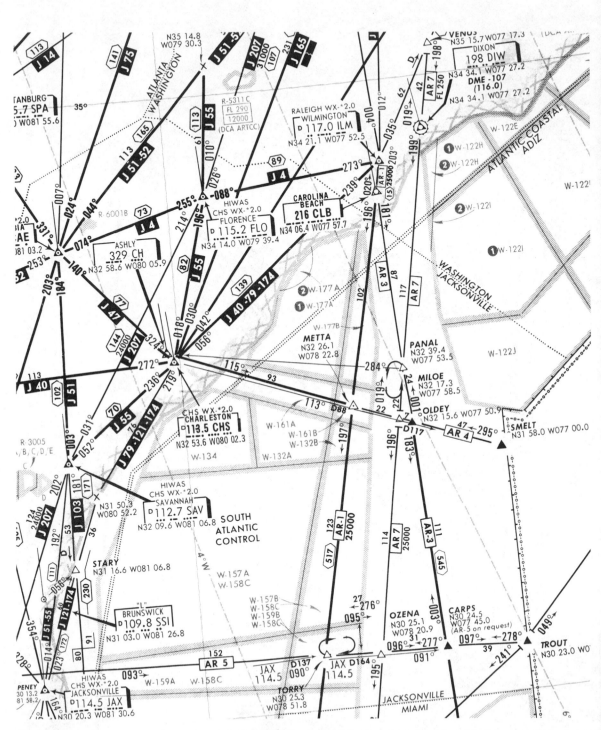

Fig. 6-5. AR-1, AR-7, and AR-3 are overwater L/MF routes connecting the US with points in Florida and the Bahamas. Reproduced with permission of Jeppesen Sanderson, Inc. Not actual size. Not to be used for navigation.

These are the major types and locations of NDBs found throughout the world. While their purposes differ, the way in which they are used is the same, and the general NDB navigational principles that follow apply to all four types.

ADF INDICATORS

The only element all three types of indicators have in common is that each points to the station at all times. It is worthwhile knowing how each type works, even if you normally only use one type.

Fixed Card

The fixed-card indicator is the simplest type of ADF indicator, which makes it both the easiest and the most difficult type to use. Since the card is fixed, zero is always at the top, and 180 is always at the bottom, and the indicated bearing to the station has no direct relationship to aircraft heading (except when the aircraft happens to be going due north), and therefore cannot directly indicate the magnetic bearing to the station. What it does indicate is Relative Bearing—the relationship of the station relative to the longitudinal axis of the aircraft. If the needle points to 90 degrees, for instance, it means the station is 90 degrees to the right of the nose—off the right wing tip; if it points to 270 degrees, it means the station is off the left wing tip; if it points to 330 degrees, it means the station is 30 degrees to the left of the nose (360 − 330 = 030), and so on.

With a fixed-card indicator, there are only two ways to determine what the actual Magnetic Bearing (the course) to the station is: one is to turn the aircraft toward the station and note the aircraft heading; the other is to add the Relative Bearing—90 degrees or 270 degrees or 330 degrees or whatever—the number of degrees indicated by the needle—to the aircraft heading. (If the result goes over 360, 360 has to be subtracted to yield a meaningful result: for instance, if the aircraft heading is 270 degrees, and the Relative Bearing is 120 degrees, then 270 + 120 = 390 degrees, or, 030 degrees Magnetic Bearing to the station.) Adding the Relative Bearing indicated on the fixed card to the aircraft heading to get the bearing to the station is fairly obvious if you think

it through, but most pilots don't want to have to think through this relationship each time they want or need to determine a Magnetic Bearing to an NDB station. Memorizing the formula:

$$MB = MH + RB$$
$$\text{where,}$$
$$MB = \text{Magnetic Bearing}$$
$$MH = \text{Magnetic Heading}$$
$$RB = \text{Relative Bearing}$$

is one way to avoid having to either turn the aircraft, or think through the relationship between Magnetic Heading and Relative Bearing, each time Magnetic Bearing is required.

A quicker way to determine Magnetic Bearing is to mentally superimpose the ADF needle on the Directional Gyro—whatever course would be indicated if the needle were actually superimposed in that position is the Magnetic Bearing To the station. This isn't very accurate, but it is a start, and it is a good double check on the reasonableness of any computations.

Rotatable Card

A rotatable card type of indicator (refer again to Figs. 6-1 and 6-2) is exactly like a fixed-card indicator, except that the card can be rotated to reflect Magnetic Heading. The advantage to being able to rotate the card is that this eliminates the need to calculate Magnetic Bearing using the MB = MH + RB formula. With Magnetic Heading set at the top of the card, the needle will automatically point to Magnetic Bearing To the station, and Magnetic Bearing From the station will be indicated by the tail. In effect, the rotatable card acts like a calculator—rotating the heading to the top of the dial automatically adds Magnetic Heading to Relative Bearing, and subtracts 360 if necessary. (This is also what happens when you mentally superimpose the ADF needle over the DG to roughly visualize Magnetic Bearing.)

The Magnetic Bearing indicated will be valid only if current heading has been set at the top of the indicator. For a continuous read-out of Magnetic Bearing to or from the station the heading must be reset each time the heading changes—even a little bit. There are times when this poses no problem, such as on a long leg with a nearly constant heading; there are other times, such as on an instrument approach when the heading is changing constantly, when it is not so practical.

Radio Magnetic Indicator—ADF

The Radio Magnetic Indicator (Fig. 4-7) solves the problems of both fixed card and rotatable card type indicators, but at considerable additional expense and complexity. It does this by providing continuous, gyro stabilized heading information to the compass card, thus providing continuously updated Magnetic Bearing information: Magnetic Bearing To the NDB station is continuously indicated under the pointer, and Magnetic Bearing From the station is continuously indicated under the tail.

Because of the differences in the way each type of indicator displays bearing information, different techniques have been developed to facilitate navigating via NDB. Except for the simple case of homing to the station, each will be examined separately in the sections that follow.

HOMING

The first NDBs were called Homing Beacons, and that is exactly what they were for: homing in on the destination. (In fact, the FAA still officially designates NDBs as *H Facilities,* presumably for that reason.) Homing is easy and instinctive, but it is also inefficient and potentially dangerous.

Since the ADF needle always points in the direction of the station (regardless of the type of indicator card), all the pilot has to do to home to a station is to point the aircraft in the direction of the station. The easiest way to do this is to turn the aircraft in the direction of the needle until the needle points to the top of the indicator. This points the aircraft directly toward the station. (See Fig. 6-6.)

Once aimed at the station, any crosswind component will displace the aircraft to one side of a direct course or the other, and the ADF needle will swing away from the top of the indicator (the nose-of-the-aircraft position), to one side of zero or the other, depending on the direction of the drift. The pilot will then have to re-aim the aircraft in the direc-

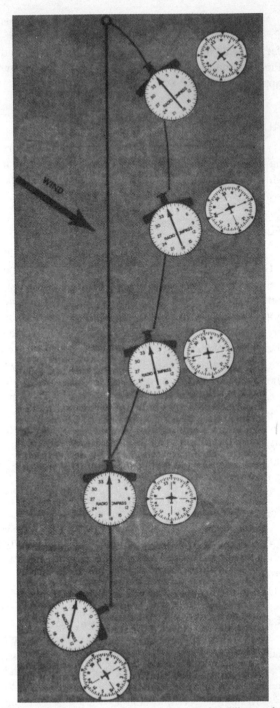

Fig. 6-6. When homing to a beacon, any crosswind will cause the aircraft to drift away from a direct course to the station, forcing the pilot to make a series of corrections to continue in the direction of the station. The result will be a curved path over the ground.

tion of the arrow to continue directly to the station. This process will have to be repeated again and again as the crosswind pushes the aircraft away from the direct course, and the resulting path to the station will be a curved one.

Repeatedly turning the aircraft in the direction of the station is not a wind correction—it is merely a wind compensation. After each turn the wind will continue to drift the aircraft away from a direct course to the station, forcing another turn, until that point where the aircraft has turned so far that it is headed directly into the wind. (Homing acts like a weathervane: the crosswind component requires the aircraft to turn further and further into the wind in order to continue toward the station, until a point is eventually reached where the aircraft is headed directly into the wind—just like a weathervane. At that point, the crosswind component is zero and the aircraft will no longer drift off a direct course, but will head straight to the station.)

The problem with homing is not so much that the resulting curved path is less efficient than a straight path (although it is), but that the actual curved path that results will be different for each crosswind/TAS combination: a strong crosswind component and a slow TAS will result in a greater curve than a weak crosswind and a fast TAS. Since the actual track over the ground will vary with every wind and airspeed combination, there is no way to ensure that any given aircraft will stay within the boundaries of an airway or approach path when homing. So while homing is a very simple procedure, and while it may have a place as an emergency technique for lost pilots, it is inefficient, and it is completely unacceptable for instrument or airway operations where the ability to stay within the confines of an airway or approach path is essential.

TRACKING

The only way to fly a straight course to the station is to *track* to the station. Tracking means to establish a wind correction angle that negates the drift caused by the crosswind. Since the ADF indicator provides only basic bearing information (as opposed to a VOR CDI which interprets radial information to command a turn to the right or left), a variety of different techniques have been developed to assist in ADF tracking. This section will examine several of these techniques as they relate to the three types of cards: fixed, rotatable, and RMI.

Fixed Card

Because a fixed card ADF indicator provides only Relative Bearing to the station, and because converting Relative Bearing to Magnetic Bearing involves arithmetic and formulas—something most pilots try to avoid in airplanes—most tracking techniques for fixed-card indicators use Relative Bearing directly to establish a wind correction angle. That is, rather than try to compute Magnetic Bearing to and from the station to see whether they are on the correct course or not, most pilots who use fixed-card indicators rely on changes in needle deflection relative to the nose or tail of the aircraft to indicate drift, course intercept, and wind correction.

The basic principle involved is a simple one: when the angle formed by the aircraft heading and the desired course is the same as the angle between either the zero or the 180 mark on the indicator and the pointer, then the aircraft is on course. (The zero mark is used to navigate To the station, and the 180 mark to navigate From the station.) For instance, if the aircraft heading is 340 degrees, and the desired course is 360 degrees inbound, then the interception angle is 20 degrees, and the aircraft will be on course when the ADF needle is 20 degrees to the right of zero—that is the point where the intercept angle and the needle deflection match. (See Fig. 6-7.) This is the equivalent of a centered needle on a CDI—not quite so obvious, but once you understand that matching needle and interception angles is the desired end, the process of obtaining it is really quite simple.

To see how this principle works in practice, assume that the desired course is 090 degrees magnetic to the beacon and that there is a direct crosswind from the left—the north. (See Fig. 6-8.) With the aircraft headed directly toward the station, the needle will initially point to the zero position (1). With a left crosswind, however, the needle will begin to move to the left of zero as the aircraft drifts to the right of course. After the aircraft has drifted several degrees off course, say 10 degrees (needle pointing 10 degrees

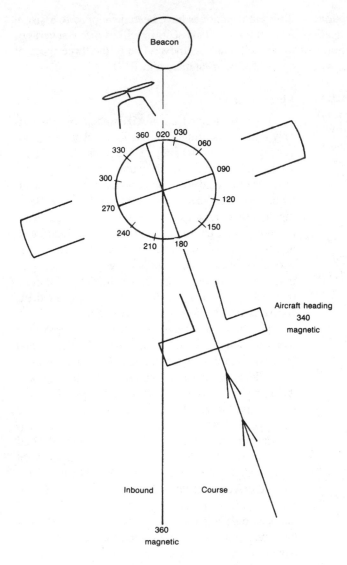

Fig. 6-7. With a magnetic heading to intercept of 340 degrees, and a desired course of 360 degrees magnetic to the beacon, the interception angle is 20 degrees. With a fixed card indicator, the aircraft will be on course when the needle deflection angle matches the interception angle—20 degrees.

to the left of zero, or 350), the need for a correction back to the 090-degree course would be apparent (2).

If you would normally use a 30-degree initial correction to track a VOR, then you would want to do the same here—30 degrees to the left, or a new heading of 060 (3). Since the needle will continue to point to the station, but the station is not in the same position relative to the aircraft heading that it was before, the ADF needle will no longer be pointing to 350, but to 020—the station hasn't changed positions, but the heading of the aircraft has. You are not yet on course at this point because the angle between the needle and the zero mark—20 degrees—does not yet match the intercept angle—30 degrees. (You can verify this with the formula, MB = MH + RB: 060 + 020 = 080, not 090, the course desired.)

With a 30-degree wind correction angle, the aircraft will almost certainly head back into the wind enough to re-intercept the original course—090. The aircraft will have re-intercepted that course when the pointer is 30 degrees (the amount of the interception angle) to the right of the zero mark—the top of the dial (4). (This can be verified again using the MB = MH + RB formula: 060 + 030 = 090.)

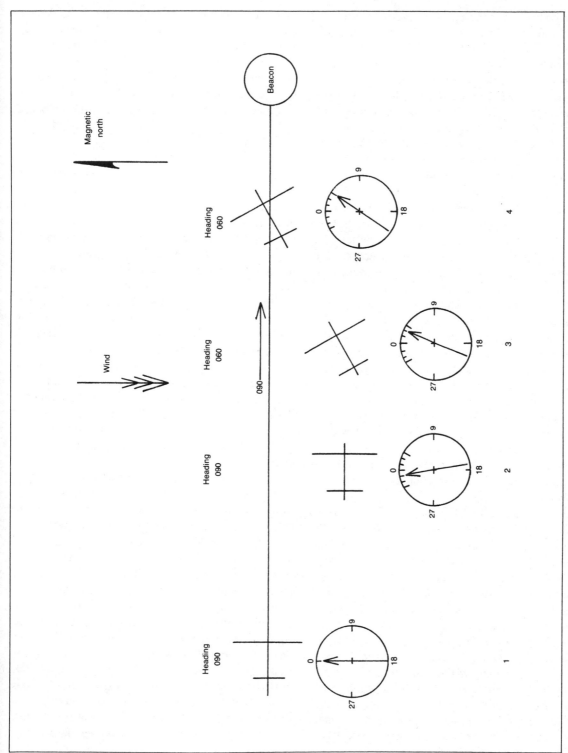

Fig. 6-8. After drifting off course to the right (2), the pilot applies a 30 degree correction to the left to re-intercept the 090 bearing to the station (3). With a fixed card indicator, the aircraft will be back on course when the ADF needle is 30 degrees to the left of the 0 mark (4).

Once back on course, the bracketing process is exactly the same to track an NDB as it is to track a VOR. A heading of 090 allowed the aircraft to drift to the right, and a heading of 060 brought the aircraft back on course, so the next heading to try is 075—halfway between the two. At this point, turning the aircraft to 075 will cause the ADF needle to be deflected 15 degrees from the zero position. As long as the needle continues to be deflected 15 degrees from the zero position (matching the 15-degree correction angle), the aircraft is on course. If it moves to the right, away from zero, then some of that correction has to be taken out—the aircraft is correcting too much and crossing over to the left side of the 090 degree bearing. If the needle moves to the left (an angle less than 15 degrees), then the correction isn't enough—the aircraft is drifting somewhat, over to the right side of the 090 bearing.

The rule here is to correct in the direction of needle movement—if the needle deflects further to the right, correct more to the right, and if the needle moves to the left (toward zero), correct more to the left. Therefore, depending on whether the needle moved from 15 degrees off the nose to either less than that or more than that, the pilot would next try something like 070 or 080, in the hope that a 20- or 10-degree correction angle would work better than the 15-degree correction angle did, and he would continue to narrow the correction angle down in this manner until a final wind correction angle had been established.

Fixed card needle deflections are directly analogous to VOR CDI indications, but less obvious: the aircraft is on the desired NDB course—the equivalent of the CDI needle centering—when the aircraft interception angle and the needle deflection angle are the same; if the needle drifts off course, the proper correction is in the direction of needle movement, which is the equivalent of correcting towards the CDI needle; the proper wind correction angle has been established when the ADF needle no longer moves—that is the equivalent of the CDI staying centered.

These rules apply equally well outbound as inbound, except that the needle deflections will be off the 180 mark—the bottom of the indicator—when navigating outbound, instead of the zero mark. Also,

as corrections are made to re-intercept the desired outbound radial, the needle will move further away from the bottom when correcting outbound, instead of crossing over to the other side of zero as it does when correcting inbound. This may seem illogical, but the only reason the needle initially moves further away from the bottom when correcting outbound is because the aircraft heading has changed; the needle will again move towards the bottom of the dial as the course is re-intercepted.

To intercept a specific course from an assigned heading with this technique, you have to know the interception angle. For instance, with an assigned heading of 220 and a clearance to intercept the 180 bearing To the station, the interception angle is 40 degrees. When the needle is 40 degrees to the left of zero, the course has been intercepted. For courses and headings that are not round numbers, such as heading 165 to intercept the 127-degree bearing outbound, it may be necessary to figure the interception angle out on paper (i.e., 165 − 127 = 38; 38 degrees is the interception angle).

If you are free to establish your own interception angle to a specific inbound or outbound bearing, the easiest way to do that is to first parallel the desired course. (See Fig. 6-9.) This gets the aircraft headed in the right direction and in a position where it is easy to orient yourself in relation to the desired course.

Once parallel to the desired course, the needle will be on one side or the other of the top or bottom of the indicator. The best way to get to the desired course from that point is to double the number of degrees the needle is deflected (up to a maximum of 90 degrees—no point in going backwards) and use that as your interception angle. This will ensure an interception angle that is directly proportional to your distance from the desired bearing. In our example (Fig. 6-9), the needle is deflected 40 degrees to the right of zero with the aircraft heading parallel to the desired course (300), so the pilot turns 80 degrees to the right to a heading of 020 (300 + 80 = 380, or 020) to intercept that course. He will know he has intercepted the desired course when the needle is 80 degrees to the left of zero—the amount of the interception angle. Once the course has been intercepted, he turns in that direction and tracks inbound to the station. This procedure works equally well to intercept a course outbound, except that the tail position (180 degrees), is used as a reference instead of the nose (or zero) position. (In fact, paralleling the desired course is a good way to get reoriented and re-established on course anytime you find yourself to be disoriented with NDB navigation.)

Rotatable Card

The main advantage of a rotatable card over a fixed card is that the rotatable card eliminates the need to do any mental calculations in order to determine Magnetic Bearing to or from the station, and Magnetic Bearing is much easier to work with than Relative Bearing: Magnetic Bearing—the desired course—is what you actually want; Relative Bearing

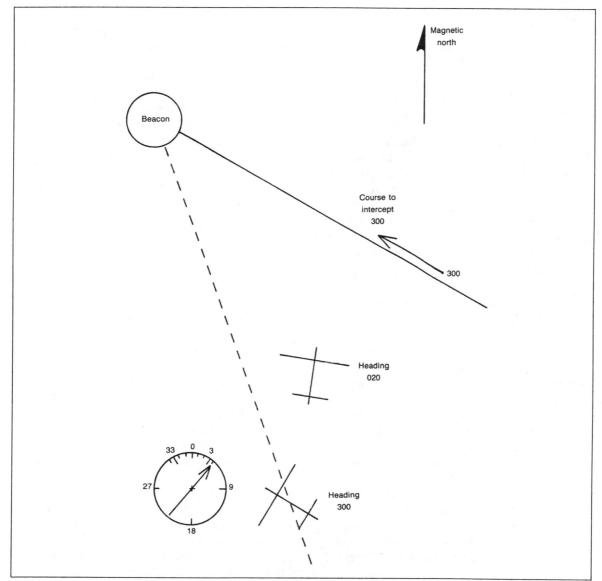

Fig. 6-9. Once parallel to the desired course, doubling the needle deflection (40 degrees in this case) will result in an appropriate interception angle—80 degrees. The heading to intercept is therefore 300 plus 80, or 020.

is just a means to an end. If the pilot has rotated the card so that present heading is under the top index, then the point of the needle will indicate Magnetic Bearing To the station, and the tail of the needle will indicate Magnetic Bearing From the station. If the Magnetic Bearing shown isn't what you want, then you need to make a correction of some sort so it is.

For example, AR-3 (Fig. 6-5) runs from the Carolina Beach NDB (CLB, 216 kHz) to the Nassau NDB and VOR (ZQA, 251 kHz and 112.7 MHz) with an outbound course (or bearing) off CLB of 181 degrees. Having observed station passage over CLB, the pilot tracking outbound on AR-3 would initially take up a heading of 181 degrees, and at the same time he would set 181 at the top of the rotatable indicator.

The chances are fairly good that at this point directly after station passage that the aircraft would not be exactly on the 181 bearing, but parallel to it on one side of the desired course or the other (just as it generally will not be exactly on the desired outbound radial after passing over a VOR). With current heading set at the top of the indicator—181—the tail of the needle will be pointing to the bearing along which the aircraft is actually flying. If the actual bearing indicated is something less than 181 degrees— say, 175 degrees—then the aircraft is to the left of course and should correct right. If the bearing indicated is greater than 181 degrees—say, 190—then the aircraft is to the right of course and should correct left.

To avoid having to think through or visualize this relationship in flight, you can rely on the rule to turn towards the pointer to intercept any course being paralleled. This rule works equally well inbound as well as outbound, except the pointer will be toward the bottom outbound, and toward the top inbound. (You can convince yourself of the validity of this rule by visualizing the aircraft parallel to the desired course, but to one side of it or the other, either inbound or outbound. Then picture where the pointer—the end with the arrow that always points to the station—would be in that situation. In every case, the proper direction to turn is toward the needle.)

The amount to turn—10 degrees, 20 degrees, 30

degrees, double the needle deflection, whatever—depends on what the pilot thinks is appropriate given his distance from the station, the expected crosswind component, and the amount he is off course at that point, just as it does when re-intercepting a VOR radial outbound after station passage. For most aircraft, a heading somewhere between 20 and 30 degrees will usually be about right.

Having taken up a heading to re-intercept, the pilot sets the new heading on the ADF card and watches the needle. With present heading set on the card, the tail of the needle will indicate Magnetic Bearing from the station, which will change, a degree at a time, as he corrects back to the desired outbound course. When the tail points to 181—the desired outbound bearing in this case—the aircraft is on course. The pilot then turns to 181 plus or minus an estimated wind correction angle, resets the new heading at the top of the indicator, and begins to track outbound.

Since the pilot is using Magnetic Bearing directly, he doesn't need to worry about matching interception angles. He is free to take up any heading that is appropriate and simply wait for the tail of the needle to point to the desired course.

Bracketing and tracking are equally simple. If the pilot can visualize aircraft position and drift from the outbound bearing indicated by the tail of the needle, then the proper direction to correct for the crosswind will be obvious. Alternately, the pilot can rely on the turn-in-the-direction-of-needle-movement rule. As long as he is careful to always set the new heading into the ADF rotatable card each time he makes a heading change, all he has to do to know when he is on course is to wait for the tail of the needle to point to the desired course.

Course intercepts, whether from an assigned heading, from a procedure turn inbound, from a terminal routing (see Chapter 8, Instrument Approaches), or at the pilot's discretion, are also easy with a rotatable card. As long as the intercept heading is set on the ADF card, the pilot simply waits for the desired outbound bearing to appear under the tail (or the pointer for an inbound bearing) to know when the desired course has been intercepted.

The rotatable card makes NDB course interception and tracking a fairly simple matter, but it does require that the ADF card be manually set to the present heading to give valid information. This is not only an additional step, but is a potential source of error—it is very easy to forget to reset a heading change, or to let the heading wander and not reset the indicator. An autopilot with heading hold can be very useful in NDB tracking in matching and maintaining aircraft heading to what has been set on the indicator.

It is also very important that whatever reference is being used for Magnetic Heading be as accurate as possible. If a non-slaved Directional Gyro is being used as the primary heading reference, for instance, it must be frequently checked against the Whiskey compass to correct for gyro precession, and the Whiskey compass itself should be as accurate as possible. To the extent the heading reference is in error, the bearing information shown by the ADF needle will also be in error by the same amount. In fact, *except when homing, NDB navigation is only as accurate as the magnetic compass to which it is referenced, regardless of the type of indicator used.*

Radio Magnetic Indicator

The Radio Magnetic Indicator, or RMI, avoids the problem of integrating aircraft heading with bearing to the station by supplying gyro stabilized heading information to the ADF indicator, thus providing the pilot with automatic, continuous magnetic bearing information. This makes the RMI much easier to use, much quicker to use, and more accurate: if the heading inadvertently wanders, the RMI will automatically change to reflect this deviation and the bearing information will still be valid. The heading input also eliminates the potential for error in setting the wrong heading into the rotatable card, or in forgetting to set it at all. (But, of course, the gyro stabilized heading information itself could be wrong, and should be frequently compared both to the primary directional gyro and the Whiskey compass.)

The RMI thus makes NDB navigation easier and safer for the pilot willing to make the additional investment. For regular NDB instrument navigation, an RMI is a virtual necessity.

Tracking with an RMI is just like tracking with a rotatable card indicator, except that the card doesn't

have to be manually rotated. (Detailed VOR tracking procedures for RMIs were covered in Chapter 4, VOR/DME Navigation; these same principles apply to NDB tracking with an RMI.) If you can visualize where the aircraft is and where it wants to be, and you keep in mind that the RMI always shows Magnetic Bearing From the station at the tail of the needle, and Magnetic Bearing To the station at the head of the needle, then navigating to or from an NDB using an RMI is a simple matter.

POSITIONING

NDBs can be used to determine present position just as VORs can. With a VOR, any given radial describes a Line of Position (LOP); using NDB, any given bearing describes an LOP. For both, the intersection of two LOPs describes a fix—present position.

Plotting LOPs with RMIs

An RMI provides Magnetic Bearing directly, which means that LOPs can be taken right off the indicator. For instance, a pilot flying along A-555 (see Fig. 6-10) could momentarily retune his ADF to the Great Inagua beacon (ZIN, 376 MHz), read the Magnetic Bearing from ZIN directly under the tail of the pointer, and use that bearing to establish an LOP. (LOPs can be plotted by aligning an instrument plotter with the Magnetic North arrow that is a part of each beacon symbol. Instructions are supplied with the plotters.) The intersection of that LOP with A-555 would indicate his present position. Alternately, the pilot could identify his position over Indee intersection (also A-555, Fig. 6-10) by observing the RMI needle: when it pointed to 169—the published bearing to ZIN over Indee—the aircraft would be over the intersection. (NDB intersection bearings are described on the charts in terms of bearing To the station, as indicated by the arrow pointing toward the station. Therefore, the point of the needle should be used to identify published NDB intersections.) For regular NDB work, dual ADFs are as essential as dual VORs are for Victor Airway navigation—one NDB for the airway, and one for position monitoring and intersection identification.

RMIs always provide Magnetic Bearing, which

works fine when plotting LOPs off NDBs where a Magnetic North reference is provided. If an LOP is being drawn using a conventional plotter against True North grid lines however, then Magnetic Bearing has to first be converted to True Bearing. (See Chapter 1, Pilotage and Dead Reckoning, for a review of this conversion.) Most instrument charts do show a Magnetic North reference for en route NDBs, and those who use charts without the Magnetic North reference (Operational Navigational Charts, Jet Navigational Charts, and other long range charts) are usually experienced military navigators or professional ferry pilots who are well aware of the distinction, but in the interest of completeness and accuracy it deserves mention: convert Magnetic Bearing to True Bearing before plotting an LOP against True North grid lines.

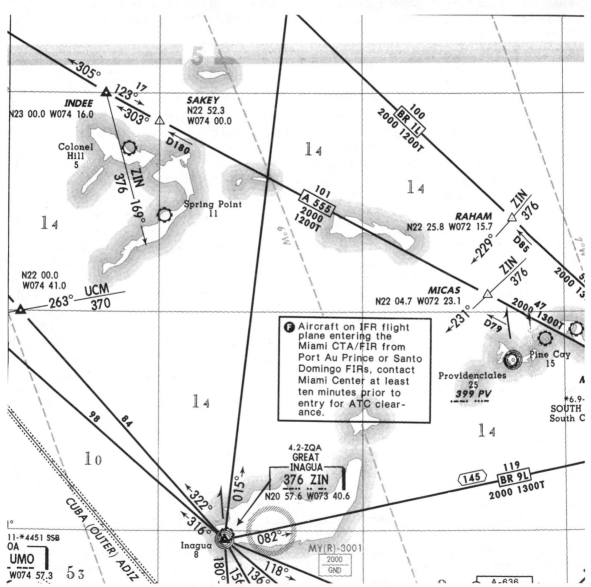

Fig. 6-10. A pilot flying along A-555 could monitor his progress along that route with an NDB tuned to the Great Inagua beacon (ZIN, 376 kHz). Reproduced with permission of Jeppesen Sanderson, Inc. Not actual size. Not to be used for navigation.

Magnetic Bearing can only be used directly with a Magnetic North arrow or compass rose that has been oriented to Magnetic North.

Plotting LOPs with Rotatable Card Indicators

The only additional step involved to plot an LOP with a rotatable card indicator (versus an RMI), is to set present heading at the top of the indicator. With present heading set at the top, Magnetic Bearing from the station can be read off the tail of the pointer and position plotted in exactly the same way as it is done with an RMI. (Magnetic Bearing to the station will be shown under the pointer itself for intersection identification.)

Plotting LOPs with Fixed Card Indicators

With the fixed card indicator, the only way to get an accurate bearing off the station is to do the arithmetic: Magnetic Heading plus Relative Bearing equals Magnetic Bearing To the station (MB = MH + RB). A pilot flying along A-555 with a fixed card indicator would have to take the aircraft heading and add it to the Relative Bearing shown on the fixed card to get Magnetic Bearing To the station (doing with arithmetic what the rotatable card does mechanically and the RMI does automatically) to identify intersections. He would then have to add or subtract 180 to that figure to get bearing From the station—the reciprocal—in order to plot an LOP. Adding the Relative Bearing to the heading and converting to the reciprocal is not difficult and is certainly not impractical for an occasional cross check of position, but I think it is easy to see why an RMI, or at least a rotatable card indicator, is desirable for those routes where NDB is essential to navigation.

LIMITATIONS

Nearly all the limitations common to NDB navigation are a direct function of its operating range in the Low and Medium Frequency (L/MF) band. The L/MF band is a convenient and practical band—nearly all early radio work was in the L/MF band and the technology is as familiar as the crystal radio set—however, as anyone who has ever listened to an AM radio at night knows, the L/MF frequency band is a band with many limitations.

Skip Zone

The signal from an NDB transmitter in the L/MF band actually has three parts: a *ground wave*, a *direct wave*, and a *sky wave*. The *ground* wave is a very stable and reliable wave, and at L/MF frequencies it may travel for several hundred miles. (The distance depends on station power and frequency—the lower the frequency and the higher the power, the stronger the ground wave. High power, over-water NDBs usually transmit at the low end of the L/MF spectrum, in order to take advantage of this phenomenon.) The *direct* wave is line-of-sight, just like a VOR signal, and can be received up to a maximum of about 200 nautical miles at the highest altitudes normally used by general aviation. The *sky* wave is a direct wave that has been refracted back to earth by the ionosphere (sometimes more than once, after having been reflected back up off the surface of the earth). The point (or points) at which the sky wave returns to earth may be many hundreds or even thousands of miles away from the source: the higher the frequency, the lower the angle the sky wave strikes the ionosphere, and the greater the distance between reception points. At a certain point—normally around the start of the Very High Frequency (VHF) band—the signal strikes the atmosphere at too low an angle and won't refract back. For this reason, the greatest radio transmission distances normally occur in the High Frequency band, just short of the VHF band—another name for High Frequency is short wave.

In most cases there will be a gap in coverage between the ground wave and the sky wave. The ground wave coverage might extend for 200 miles, for instance, with a sky wave returning to earth at 1,000 miles; in this case, nothing would be received between 200 and 1000 miles. This gap is called the Skip Zone. The exact length of the skip zone depends on frequency, power and atmosphere, and is hard to predict. This variability limits the effectiveness and reliability of NDB navigation.

Night Effect

Light passing through the atmosphere determines

the height of the ionosphere, which in turn determines the length of the skip zone. This causes special problems at dawn and dusk when the ionosphere is changing from one position to another. At certain times the ground and sky waves may overlap, and the ADF receiver may have difficulty differentiating between the two signals, resulting in temporarily erratic indications. This is called *Night Effect*. It will usually resolve itself once the sun has fully risen or set.

Fading

Fading of the signal occurs mainly at night. It is caused by the simultaneous reception of both the ground and sky waves, or sometimes by two sky waves. Since the ground and sky waves travel different distances, they arrive at the receiving antenna at different times, sometimes *in-phase*, meaning the wave patterns happen to coincide, and sometimes *out-of-phase*, meaning the wave patterns conflict. Out-of-phase signals tend to cancel each other, while in-phase signals reinforce each other. Since the atmosphere is actually changing all the time, the two signals are constantly going in and out of phase, resulting in a rising and falling signal. The normal cockpit indication of this condition is a rhythmical swinging of the needle, with a corresponding building and fading of the identification signal. When this happens, the pilot is forced to navigate as best he can by averaging the needle swings, and by attempting to note the bearing when the ident signal seems strongest.

Shore Line Effect

The ground wave travels in one direction over land, but another over water—the difference in conductivity between the two causes the signal to be refracted, or bent. This is called *Shore Line Effect*, and results in erroneous bearing indications. Most over-water NDBs are situated as close to the water as possible in order to minimize this error. NDB airways, such as AR-3, have already allowed for any bending, but pilots taking bearings over water from NDBs situated inland should be aware of this error, and should not attribute too much accuracy to an LOP established off inland beacons.

Terrain

Other terrain factors which affect bearing indications are mountains and natural magnetic disturbances. NDB airways take these errors into account as best they can, and areas of known magnetic disturbance are marked on the appropriate charts, but these represent additional sources of inaccuracy in NDB navigation.

Interference

When two VOR signals with the same frequency are received, the VOR receiver tends to select the strongest one and reject the other. An ADF receiver, on the other hand, tends to alternate back and forth between the two signals until one is dominant enough to hold the signal. A regular swinging of the needle from one bearing to another is the usual symptom of interference, and can be verified by the reception of two idents. It will usually resolve itself as the aircraft moves closer to one station and away from the other.

Thunderstorms

Lightning creates electrical disturbances in the L/MF band. These disturbances take the form of static interference to the audio ident, and erroneous or erratic needle indications. (In fact, the ADF can be used as a very crude lightning indicator, since the needle will sometimes point in the direction of the disturbance.) For all practical purposes, the ADF is unusable in and around areas of thunderstorm activity. This can be a serious limitation if the pilot is attempting to use NDB to locate an airport in order to land and avoid the thunderstorm activity causing the interference. There is no solution to this problem other than prevention.

ACCURACY

Non-Directional Beacons are subject to so many different factors in their signal propagation characteristics that it is impossible to specify an absolute accuracy standard for NDB; however, the cumulative effect of all errors (including whatever error is inherent in the compass reference itself) is a considerable degree of inaccuracy—it is impossible to be any more specific than that. The greatest difficulty with NDB however, is the variability of its accuracy—it is virtually impossible to know what the actual accuracy is at any given moment beyond a rough guess. Nonetheless, within the nominal range limitation for each type of NDB—15 nautical miles for Compass Locators, 25 nautical miles for approach aids, 50 nautical miles for en route beacons, and 75 nautical miles for high power beacons—NDB accuracy is adequate for the purpose intended, and it may be adequate, at times, beyond those range limitations.

CONCLUSION

NDB is not a terribly accurate system of navigation: that is exactly the reason VOR was developed. It is, however, much better than no system at all, especially in light of its low cost. NDB should always be thought of as a supplement to other forms of electronic navigation when they are available, and when NDB is the only navigational information available the pilot should be careful not to ascribe a degree of accuracy to it for which it is not capable. As an approach aid, NDB is capable of little more than a controlled descent in a safe area to what the pilot hopes will be visual conditions; as a primary en route navigational aid, it should be supplemented by dead reckoning estimates and headings, and should not be relied upon unless nothing else is available; as an emergency navigational aid, it can provide an approximate indication of position and serve as a homing point. As long as these limitations are kept in mind, NDB can be a useful, and at times essential, navigational aid.

Chapter 7

Ground-Based Radar Navigation

Radar (radio detection and ranging) was developed for the military just prior to World War II. Since that time, radar has proven to be one of the most important developments ever in aviation technology. The military continues to rely on radar, both ground based and airborne, to a much greater extent than civil aviation does, but ground based radar is nonetheless still extremely important in the control, separation, and navigation of civil aircraft as well.

If you are curious about some of the unclassified military uses for radar, you might want to read *Air Navigation*: Flying Training, AF Manual 51-40/NAVAIR 00-80V-49. It is available at many libraries, or you can order it from the Superintendent of Documents (see Appendix A). This Air Force/Navy training manual describes some of the more esoteric uses of radar: low level navigation, aerial refueling, blind formation flying, and radar-based dead reckoning. Our concern in this chapter is not with these specialized uses for radar however; our concern here is with the use of ground-based radar to facilitate conventional air navigation and control.

THE ROLE OF RADAR IN AIR NAVIGATION

One way to look at the role ground-based radar plays in the overall air navigation picture is to divide air navigation into three major categories. The first category would be navigation based on external aids: VOR, NDB, DME, LORAN-C (Chapter 9), OMEGA/VLF (Chapter 10), a couple of systems that are now obsolete (CONSOLAN and Four Course Ranges), and a satellite system that is not yet operational (NAVSTAR/GPS—Chapter 12). All of these systems are dependent on electronic aids to navigation outside the aircraft.

The second category would be self-contained navigation: pilotage, dead reckoning, inertial navigation (Chapter 11), celestial navigation (not covered in this book) and a somewhat obsolete system called Doppler Radar (also not covered in this book). (Both Doppler and celestial nav are covered in the *Air Navigation* manual previously cited.) These navigational techniques can be conducted from the cockpit without any external navigational aids.

The last category would be ground controlled

navigation—navigation which requires neither external aids nor self-contained systems. As far as current technology is concerned, this means radar, and that is the subject of this chapter.

RADAR PRINCIPLES

While the actual operation of radar is enormously complicated, all radar, whether ground-based or airborne and regardless of whether its primary function is aircraft surveillance, ground mapping, or weather detection, is based on a simple principle: electromagnetic energy radiating outward from a source will be reflected back toward that source by objects in its path. Since electromagnetic energy travels at the speed of light (a constant speed for all practical purposes), the amount of time it takes a pulse of energy to reach an object and return can be used directly to determine the distance to that object—the greater the amount of time, the farther the object is away. The position of the antenna at the moment the energy is received can be used to indicate the azimuth and elevation of the object in relation to the radar. Distance, azimuth, and elevation can then be used to plot the object on a plan view of the area covered by the radar.

The principle is simple, but the execution is complex. Timing is very critical—an error of as little as a single microsecond (a millionth of a second) will result in an error in distance of almost 500 feet. The accuracy of the distance measurement is therefore only as accurate as the clock that times the interval between transmission and receipt of the radar energy. Timing is also important in determining azimuth and elevation—the radar receiver must correlate the exact moment the return is received with antenna position to obtain azimuth and elevation to the object. The overall accuracy of the point plotted then, both in terms of distance and in terms of direction and elevation from the radar antenna, is directly related to the accuracy of the timing device used.

An even greater difficulty, however, is that enormous amounts of energy must be transmitted in order for even a tiny bit of energy to be returned. This means that the radar has to be both very powerful as a transmitter and very sensitive as a receiver, and the two are not compatible—the transmitter tends to completely overpower the receiver. The only practical solution is to alternate transmitting and receiving at very brief intervals and at very high rates. (High rates are necessary in order to create an accurate and clear picture, just as a high shutter speed is necessary to capture a rapidly moving object photographically). At a typical range of 40 miles for instance, the radar will cycle between transmitting and receiving approximately 800 times per second.

These are difficult and complex problems, but they have been overcome largely through some very tricky engineering. With the solution comes a price though, both literally and figuratively: radars are expensive, and they require careful tuning and adjustment of all the various components to function properly (or at all). They also tend to be large, heavy units drawing huge amounts of power, generating lots of heat, and requiring constant maintenance to perform in peak condition.

There are other difficulties, too. The only practical way to create large amounts of energy at very short intervals—pulses—is with microwaves, and microwaves, like all electromagnetic energy above the High Frequency band, travel only in a line of sight. This limits range and causes problems with terrain masking. Microwave energy is also vulnerable to refraction, or bending, by temperature inversions in the atmosphere (temperature increasing with altitude). If the beam of energy is bent toward the ground, it can cause extraneous targets to appear on the screen; if it is bent upward, it disappears into space and no return is generated at all. Radars designed specifically for aircraft surveillance have circuits to eliminate reflections off objects on the ground, which helps with the problem of extraneous targets, but little can be done about bending upwards.

Radar also isn't very smart, despite computer processing of much of the information. It isn't very good at all, for instance, at differentiating among the various objects which it strikes. Because of this, ground-based radars that want to see only airplanes also see precipitation, and airborne radar that wants to see only precipitation also sees terrain.

SECONDARY VERSES PRIMARY RADAR

Because of these many limitations, radar control of aircraft based on the primary return of energy from the aircraft being controlled poses many problems: the energy returned is very often too low to be detected by the radar receiver; ground clutter and precipitation often mask some aircraft returns; even when a return is strong enough to be detected, there is no way to tell one return from another except to ask the pilot where he is and look for a return in that area (which results in the interesting situation where the pilot tells the controller where he is, so the controller can tell the pilot where he is). An alternate method of identification is to have the aircraft make an identifying turn; the radar return that responds is the correct one. Neither method is very fast, and neither method is fail safe.

Many of these problems are avoided with a device called a *radar transponder*. A radar transponder is a small microwave receiver and transmitter that generates an artificial radar return whenever it senses a radar transmission, or "interrogation." (See Fig. 7-1.) The transponder-reply signal is much more powerful than the reflected energy of the original radar transmission itself, providing the ground-based radar receiver with a much higher level of energy with which to work. This essentially solves the problem of weak returns and eliminates some of the problems with terrain masking and precipitation interference. In addition, the transponder reply can be coded, which provides a much more positive way to identify aircraft than the two methods described previously. (The transponder in Fig. 7-1, for in-

stance, is set to respond on code 2412.) For an even more positive identification, the controller can ask the pilot for an *ident*—a cockpit initiated identification signal accomplished by pushing the button marked "Ident" Fig. 7-1. The Ident function causes the return from that particular transponder to glow brightly on the radar screen, unmistakably identifying the aircraft in question to the controller.

When the image displayed on the controller's scope is a transponder-produced image, the return is called a *Secondary Return*, and the system which decodes and displays it is called a *Secondary Radar System*. This is because the image displayed is not a primary image based on a direct reflection of energy off the aircraft, but a secondary response to a radar interrogation. Figure 7-2 is a drawing taken from the *AIM* which illustrates one type of radar controller's screen, with an explanation of what the various types of returns mean.

Most of the radar systems used by the FAA also include a variety of supporting information with each aircraft target symbol: aircraft identification, altitude, ground speed, vertical trend, projected flight path, and sometimes runway assignment or arrival airport. In this way, the radar controller has available to him all the information he needs to control aircraft and to provide other non-essential radar services.

The images produced on the controller's screen are a confusing glow of symbols to a non-controller, but to the controller they represent a clear picture of aircraft under his control and their relationship to each other, to terrain features, to navigational aids, and to other aircraft in the area not in communication with the controller. The best way to see what an ATC radar display actually looks like (and also the best way to understand how the system works), is to visit an approach control or en route facility, and pilots are encouraged by the FAA to do so. Your local General Aviation District Office (GADO) can help you in arranging a visit; Flight Service can usually help with phone numbers if you wish to contact a nearby approach control or en route center directly.

TRAFFIC ADVISORIES

The air traffic controller normally sees all transponder-equipped aircraft in his area of cover-

Fig. 7-1. Transponders are used to provide positive, coded replies to ground based radar interrogations. Photo courtesy of Narco Avionics.

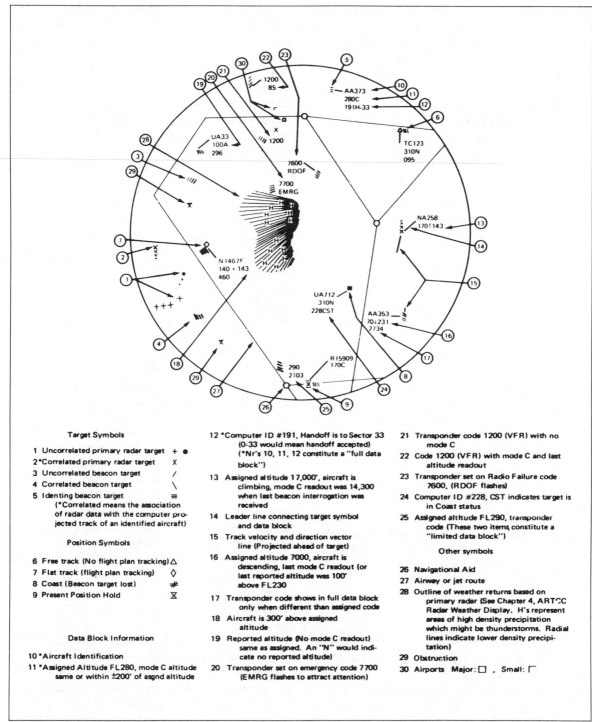

Fig. 7-2. Modern radar systems display much more than primary returns from aircraft, as this drawing, with accompanying legend (taken from the *Airman's Information Manual*), shows.

Target Symbols

1. Uncorrelated primary radar target   + ●
2. *Correlated primary radar target   X
3. Uncorrelated beacon target   /
4. Correlated beacon target   \
5. Identing beacon target   ≡
   (*Correlated means the association of radar data with the computer projected track of an identified aircraft)

Position Symbols

6. Free track (No flight plan tracking) △
7. Flat track (flight plan tracking) ◇
8. Coast (Beacon target lost) ⚹
9. Present Position Hold ⊠

Data Block Information

10. *Aircraft Identification
11. *Assigned Altitude FL280, mode C altitude same or within ±200' of asgnd altitude

12. *Computer ID #191, Handoff is to Sector 33 (0-33 would mean handoff accepted) (*Nr's 10, 11, 12 constitute a "full data block")
13. Assigned altitude 17,000', aircraft is climbing, mode C readout was 14,300 when last beacon interrogation was received
14. Leader line connecting target symbol and data block
15. Track velocity and direction vector line (Projected ahead of target)
16. Assigned altitude 7000, aircraft is descending, last mode C readout (or last reported altitude was 100' above FL230
17. Transponder code shows in full data block only when different than assigned code
18. Aircraft is 300' above assigned altitude
19. Reported altitude (No mode C readout) same as assigned. An "N" would indicate no reported altitude)
20. Transponder set on emergency code 7700 (EMRG flashes to attract attention)

21. Transponder code 1200 (VFR) with no mode C
22. Code 1200 (VFR) with mode C and last altitude readout
23. Transponder set on Radio Failure code 7600, (RDOF flashes)
24. Computer ID #228, CST indicates target is in Coast status
25. Assigned altitude FL290, transponder code (These two items constitute a "limited data block")

Other symbols

26. Navigational Aid
27. Airway or jet route
28. Outline of weather returns based on primary radar (See Chapter 4, ARTCC Radar Weather Display. H's represent areas of high density precipitation which might be thunderstorms. Radial lines indicate lower density precipitation)
29. Obstruction
30. Airports Major: □ , Small: ⌐

101

age, and will also see the primary returns of some of the non-transponder-equipped aircraft in his area of coverage. Workload permitting, he can then advise those pilots with whom he is in contact of potentially conflicting traffic. Traffic advisories are not his primary mission though—aircraft control is his primary mission. The issuance of traffic advisories is not considered an essential, or primary service, because pilots operating in visual conditions have primary responsibility to see and avoid conflicting traffic for themselves, and pilots operating IFR are provided separation from other IFR traffic directly by ATC. Traffic advisories provide an additional measure of safety—a back-up to the primary VFR and IFR systems for aircraft separation.

VECTORS

In aviation, a *vector* is an assigned heading—a direction to fly. Controllers assign vectors for aircraft separation, and to facilitate other forms of air navigation.

When a controller issues a vector, he assumes primary responsibility for the navigation of the flight. This does not relieve the pilot of his ultimate responsibility for the safety of the flight—the pilot must still monitor his position to ensure that the controller does not issue unsafe vectors and to be able to resume his own navigation at some later point—but it does relieve him of the immediate responsibility for navigation.

Vectors have many advantages both to the pilot and the controller, and are an essential part of the normal air navigational procedures in any high density environment (such as around any major airport or airway). The following sections examine several of these applications individually.

Separation

One of the primary uses for vectors is in aircraft separation. When the anticipated flight paths of two aircraft appear to be in conflict, rather than change the routing of one of the aircraft, or cause one to hold in position while the other passes, the controller can issue a vector to one of the pilots so that the flight paths no longer conflict. This is generally prefera-

ble to a change in the routing, and is nearly always preferable to holding. (The controller can also issue a change in the airspeed of one or both of the aircraft.)

Having issued a change in either the heading or airspeed, the controller then monitors his scope to ensure that the flight path corrections result in the separation desired, and issues further vectors or speed changes if they do not. Once the aircraft have passed each other, the controller issues a new clearance to return the vectored aircraft back on course or back to normal airspeed.

Sequencing

When aircraft converge on a point, the controller must sequence them to avoid conflicts. Vectors are a common way of doing this. If, for instance, all aircraft arriving at an airport must pass over the outer marker in order to complete an ILS approach, the controller can issue vectors as necessary to ensure that the arrival times are staggered and that the arriving aircraft are *in trail*—a line of aircraft with appropriate separation between each other. This is more efficient, and generally also safer, than the alternative of having aircraft hold at different altitudes over the marker, working each down a thousand feet at a time as the lowest aircraft in the stack breaks off to complete the approach. (In bad weather though, with long delays, stacking is still sometimes necessary.)

Radar sequencing is also commonly used en route to properly space all aircraft along a particular route. A faster aircraft that is gaining on a slower aircraft, for instance, could be issued a vector first off of, and then parallel to the airway. Once past the slower aircraft, another vector would then be issued to rejoin the airway. The controller is generally in the best position to determine whether the vector off the airway is preferable to a speed reduction, although the pilot can always ask for a speed reduction instead of a vector if he thinks that would be preferable. The final decision lies with the controller however. (The pilot always has the option to refuse a clearance, but he should have a very good reason for doing so—refusing a clearance is disruptive and may require the controller to hold the aircraft in position until the pilot's refusal can be accommodated.)

Navigational Assistance

There are a variety of ways in which the controller can use vectors to assist the pilot with his own navigation. One of the more common is to issue vectors to intercept the inbound course on an instrument approach. Vectors reduce the distance that must be flown to establish the aircraft on the inbound course, simplifying the approach procedure and reducing the pilot's workload. Radar vectors also speed up the arrival process for all aircraft. (For more information on instrument approaches, see Chapter 8.)

Vectors can also be used to intercept specific courses and airways. An aircraft departing Hilton Head Airport (Hilton Head, South Carolina—see Fig. 7-3), with a clearance to fly a heading of 030 to intercept V-437 northbound (the 052 radial off the Savannah VOR) would be able to proceed more or less directly on course instead of having to navigate directly to the Savannah VOR first.

Direct Courses

Pilots navigating using VOR/DME guidance generally must overfly the associated VOR/DME facilities in order to obtain directional guidance. (Refer back to Chapter 4, VOR/DME Navigation, for more on Airway navigation.) This adds to the distance flown and increases the pilot workload somewhat. Radar can frequently be used in place of VOR/DME navigation by providing clearances along a direct course. En route air traffic control centers are able to do this by combining the information from several radars, which has the effect of creating continuous radar coverage for hundreds of miles—much greater than the range of a single VOR. With this overview, radar controllers can frequently provide vectors towards a facility that would otherwise be well beyond the normal VOR reception range. A pilot taking off from Tallahassee for Nashville via J-41 to intercept J-39, for instance, (see Fig. 7-4), might be able to get a radar vector after takeoff from Tallahassee direct to Nashville, and in so doing save 29 nautical miles (about eight percent of the airway distance) and reduce his navigational workload. During busy times, vectors for points well ahead of the

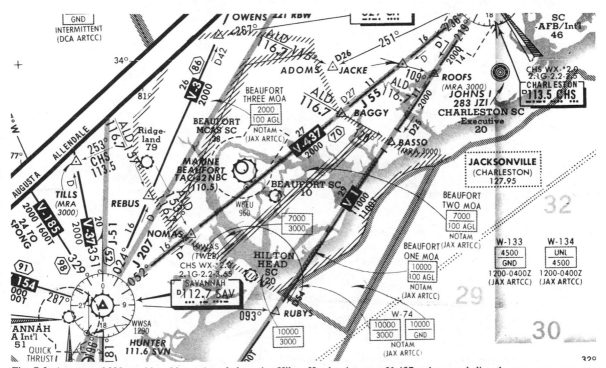

Fig. 7-3. A vector of 030 would enable an aircraft departing Hilton Head to intercept V-437 and proceed directly on course en route to Charleston. Reproduced with permission of Jeppesen Sanderson, Inc. Not actual size. Not to be used for navigation.

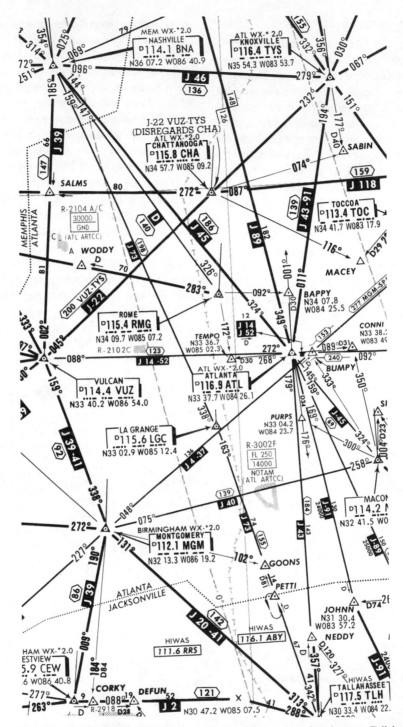

Fig. 7-4. Radar controllers can sometimes provide vectors between facilities far apart, such as between Tallahassee and Nashville (traffic permitting). This shortens the en route distance and reduces the pilot's workload. Reproduced with permission of Jeppesen Sanderson, Inc. Not actual size. Not to be used for navigation.

aircraft are not always possible, but when traffic is light (such as late at night) vectors for points many hundreds of miles ahead are not at all uncommon.

Vectors for Visual Approaches

Radar vectors can be used to locate the destination airport visually. It is a fairly simple matter for the controller to provide the pilot with a vector towards his destination airport (and, in some cases, even to specific runways), since most airports and some runways are plotted on his screen. It is still the pilot's responsibility to find and identify the proper airport, but the vector can be a big help in doing so. If the airport is located visually, then the pilot can be cleared for a visual approach, eliminating the need for a time and fuel consuming instrument approach.

Emergency Assistance

Radar vectors can also be used to help the VFR pilot who inadvertently flies into the clouds find the shortest route back to visual conditions. By combining information from other aircraft, from precipitation displayed on his screen, and from current weather reports, the radar controller can provide non-instrument rated pilots with a vector that seems most likely to provide the shortest route out of instrument conditions, or to low terrain with high ceilings where a controlled let-down can be accomplished.

This capability can also be used by any pilot with a problem en route to locate the nearest airport or best emergency landing area. An aircraft with a heavy load of ice, for instance, could request a vector to the nearest airport with long runways, or a single engine aircraft experiencing engine problems at night could request an immediate vector to the nearest lighted airport.

Many more examples could be given to illustrate the ways in which radar vectors can be used to facilitate air navigation and increase the overall level of safety, but these examples cover the most common areas. Radar vectors—the assignment of headings based on radar observations—are the most fundamental of all ground based radar navigational procedures.

RESUMING NORMAL NAVIGATION

Vectors are, by nature, expedients. Once their purpose has been served, they are superseded by fur-

ther clearances to specific fixes: "Proceed direct the ABC VOR. Resume own navigation," for instance. The controller will sometimes include the point where the resumption of normal navigation is expected when the vector is issued—"Fly heading 030 to intercept the ABC 090 radial, resume own navigation,"—but more often than not the controller will wait until the appropriate fix is being approached to cancel the vector and return the pilot to his normal routing.

Controllers will always inform the pilot of the purpose of the vector with the initial clearance: "Vectors for aircraft separation"; "Vectors for the airport"; "Vectors to intercept Victor Three," and so on. If radio contact is lost, the pilot is then expected to proceed, by the most direct route, to the fix specified. (In the case of vectors for aircraft separation, this would mean back to the original routing.)

POSITIONING

Ground-based radar can be used to determine present position just like any other form of navigation. The radar screen is calibrated in such a way that the controller can tell the pilot where he is either in terms of distance from a fix ("You are presently three miles from the outer marker, cleared the ILS 23 approach"), radial and distance from a VOR ("Present position is the ABC 090 radial, 23 DME"), or latitude and longitude ("I show you North 39 38.0, West 069 46.4").

Radar positioning can also be used in place of other ground-based navigational facilities and fixes. For instance, if the ILS outer marker is inoperative (or the aircraft marker beacon receiver is inoperative), the controller can "call-out" the outer marker based on his radar observation of the aircraft position. Radar positioning can also be used as a substitute for DME fixes, and for airway intersections.

GROUND SPEED DETERMINATION

The radar system normally used by ATC (the FAA calls it the ARTS III radar system) is a computerized system. The computer keeps track of all the data that goes with each target, projects flight paths, and, as a necessary part of those projections, computes and displays an average ground speed for each target. Therefore, even if you do not have DME onboard, you can still obtain a ground speed estimate simply by asking the controller. This estimate will normally not be as accurate as one computed by the pilot, but it is good to know that this information is available when you have been unable, for one reason or another, to determine an accurate ground speed estimate for yourself. It can also be a good double check for those times when the ground speed estimate that you have computed seems to be unreasonable, or to vary considerably from what was expected.

SURVEILLANCE

Radar is a worthwhile system for its surveillance ability alone—its ability to monitor normal navigation and prevent errors. It is very easy to set the wrong radial in a VOR CDI for instance, and fly off the airway. With radar, this kind of deviation would be quickly detected and corrected. It is also very easy to miss a turn point, to fail to intercept an NDB bearing, to program an RNAV computer incorrectly, or to enter the wrong lat/long coordinates into a long range nav system (Chapter 9). In each case, without radar the error might not be apparent until a serious deviation had occurred, but with radar the deviation would be immediately apparent to the controller. Radar is not a substitute for careful navigation, but it is an important back-up.

CONCLUSION

Radar has many limitations, but it is still an enormously useful method of aircraft control and a very effective expedient in air navigation. Transponders have eliminated most of the more serious limitations of radar, and computers have enabled it to be much more than the simple "radio detection and ranging" device that it was originally. Radar is not, in theory, absolutely necessary for either navigation or aircraft control and separation, but it has so facilitated air travel that it has become a virtual necessity in the more heavily traveled areas of the world. Its most important role though, may be as a very effective safety device, monitoring the movement of aircraft, catching navigational errors before they become serious, and providing the pilot with an additional set of eyes in visual conditions and positive aircraft separation in instrument conditions. Despite occasional complaints that the U.S. system of air traffic control and separation is too radar oriented (i.e., too ground oriented), the navigational advantages of radar are considerable, and no better system of aircraft control currently exists. Accordingly, ground-based radar surveillance and control can almost certainly be expected to grow in importance and in coverage in the years ahead.

Chapter 8

Instrument Approaches

Air navigation can be divided into two major segments: an en route segment, and an approach segment. The en route segment is large scale navigation: accurate estimates of time and fuel are very important here, precise position estimates in relation to the terrain generally are not. The approach segment is small scale navigation: precise knowledge of position is a major factor here, highly accurate time and fuel remaining estimates generally are not. During the approach phase, the proximity to terrain forces a reversal of the normal navigational priorities.

We could call en route navigation Macro Navigation: large scale, relatively coarse navigation. Approach navigation would then be Micro Navigation: small scale, precise navigation. If macro navigation paints large canvases with a broad, flat brush, micro navigation paints miniatures with a fine, pointed brush. This chapter is about painting miniatures, with airplanes.

In aviation, the word approach can have many meanings. In this chapter, when we talk about approaches we mean instrument approaches: navigation under Instrument Meteorological Conditions (IMC)

to Visual Meteorological Conditions (VMC), for the purpose of establishing visual contact with an airport or runway in order to complete a safe landing.

IMC is a legal term. IMC exists whenever the visibility is less than that required for VMC, for that particular situation. IMC is therefore not an absolute, but depends upon the type of airspace, the type of aircraft, the aircraft altitude, and the time of day.

In practice, IMC means those conditions where the visibility is limited due to clouds, rain, fog, snow, smoke, haze, drizzle or any other restriction to visibility such that adequate ground contact cannot be maintained for the purpose of navigating via pilotage and dead reckoning. IMC is commonly thought of as being "in the clouds," but it is not necessary to be actually in the clouds to be in IMC—any restriction to visibility qualifies. Under these conditions, navigation must be done by instruments—electronic aids to navigation—by a pilot with an Instrument Rating, and the aircraft must be operated on an IFR flight plan and have an appropriate IFR clearance.

Pilots without Instrument Ratings (and who are

therefore prohibited from filing IFR flight plans and operating on instrument clearances), can still benefit from this chapter. No regulation prohibits non-instrument-rated pilots from flying instrument approaches, as long as they do them under visual conditions and observe the usual VFR restrictions and regulations. The advantages provided by instrument approaches in safely and positively locating and identifying desired airports and runways are substantial and are not limited to effecting transitions from instrument to visual conditions. VFR pilots who have become frustrated by difficult searches for airports at the ends of their flights—frustrated despite careful flight planning and experience with pilotage, dead reckoning, and basic VOR navigation—will find a knowledge of instrument approaches to be very helpful in navigating to a precise point.

This chapter is not meant to be a substitute for instrument training in approach procedures, nor is it intended to be a substitute for careful study of chart legend information, introductory material supplied by the approach chart manufacturers, or the *Airman's Information Manual* (AIM). The emphasis in this chapter is on the navigational aspects of instrument approach procedures, and no attempt will be made to cover every detail of approach chart interpretation or the specific techniques for flying instrument approach procedures. (The practical aspects of flying instrument approaches are covered in my previous book, *Fly Like A Pro*, TAB book #2378.)

ELEMENTS COMMON TO ALL INSTRUMENT APPROACHES

There are many different types of instrument approaches, each with different capabilities, characteristics, and limitations. All approaches nonetheless have certain elements in common, and the ways in which approaches are alike are worth examining before looking at the ways in which they differ.

Unless otherwise indicated, all numbers in parenthesis that follow refer to numbered references depicted in Fig. 8-1, which is an illustration of a typical approach procedure, in chart form. The approach chart used is a Jeppesen approach chart—Jeppesen is the leading supplier of commercial approach charts as well as commercial en route charts for civil avia-

tion. NOS (National Ocean Service), or "government" approach charts are also available (Appendix B), and are identical in content to Jeppesen charts, differing only in symbol style, layout, and renewal format from Jeppesens.

Prescribed Area

Every approach has a prescribed area for maneuvering in which certain restrictions apply. This area is normally 10 nautical miles from the primary approach fix, as shown by the note indicated at (1), but the prescribed area can be as small as five nautical miles and as large as 15 nautical miles.

Primary Elements

All air navigation can be broken down into three primary elements: range, azimuth, and elevation. (Pilots commonly think of these three elements as distance, direction, and altitude.) These three primary elements also describe the essential navigational components of any approach procedure.

Range. Range, or distance to a fix, is not as important in approach navigation as it is in en route navigation—azimuth and elevation are much more important. Range is still important though, and ways are provided in every approach for the pilot to determine his distance from key points in the approach and to help him in identifying his position in relation to the airport.

The best range information is provided by DME, however DME is not always available. Normally therefore, aircraft position is indicated at various points during the approach by the passage of published intersections, nav aids, and marker beacons. Distance from these fixes to other points on the approach and to the airport itself is shown on the profile view portion of the approach chart (2). In our example (Fig. 8-1), the only range marking available is from the McAlester VOR itself (MLC, 112.0 MHz) to a point over the runway: 1.5 nautical miles.

Azimuth. Azimuth, or direction, is critical to all instrument approaches. The approach course is normally a single direction, and is based on courses, radials or bearings from the primary nav aid. (A very few approaches have two courses, one before the pri-

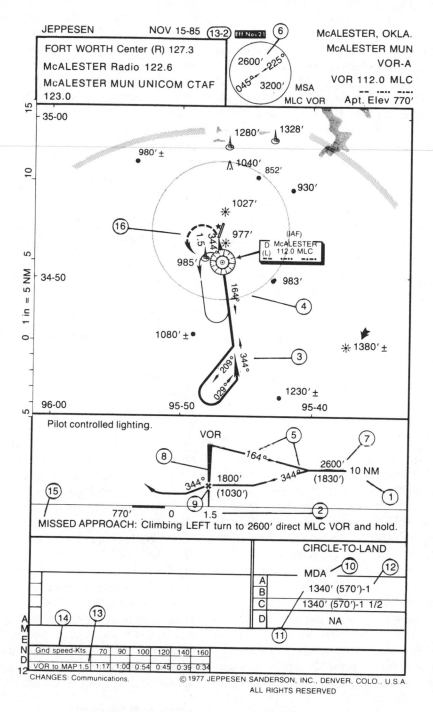

Fig. 8-1. Standard instrument approach charts are used to describe individual approaches. Numbered items refer to text. Reproduced with permission of Jeppesen Sanderson, Inc. Not actual size. Not to be used for navigation.

mary fix, and a slightly different one after. These are called dogleg approaches; they bear careful watching, as it is very easy to forget to alter course after station passage.) The primary inbound course for the McAlester approach is the 344-degree radial To the MLC VOR (3) (which is also the 164-degree radial From the VOR [4]). The outbound radial from the VOR to the airport is also 344 degrees—this is not a dogleg approach. (The outbound course is hard to see on this approach, because the distance from the VOR to the airport is short and there is very little room to print it on the chart, however, if you look carefully you can see the outbound course, 344, between the VOR symbol and the runway symbol, along with the distance—1.5 nautical miles.)

In any approach, it is imperative that the correct course be flown. Terrain clearance can be assured only when the aircraft is properly located on the prescribed course. Course is shown most clearly on the plan view, but it is also shown on the profile view (5).

Elevation. Elevation, or altitude, is the most critical element in any approach procedure. Every point on every approach has a minimum altitude below which terrain and obstacle clearance and proper nav aid reception are no longer assured. Minimum altitudes are always shown on the approach chart itself—the controller will not supply these altitudes.

For the general area surrounding the primary approach fix, *Minimum Sector Altitudes* (MSA) apply. These altitudes are for emergency use only and guarantee 1,000 feet of terrain clearance within 25 nautical miles of the primary facility, but do not guarantee nav aid reception. (MSAs are the approach equivalent of MOCAs for airways.) Minimum Sector Altitudes are shown on Jeppesen charts in the form of a small diagram at the top of the chart (6). For this approach, the Minimum Sector Altitude for the area to the north and west between the 045 and 225 degree radials off the McAlester VOR is 2600 feet MSL; the MSA for the sector to the south and east of those radials is 3,200 feet MSL. Approaches may have as many as four different sector altitudes, depending on the terrain.

Normal (non-emergency) minimum altitudes for the various segments of the approach are shown on the profile view—this is, in fact, its primary function. The profile view for this approach shows a minimum altitude of 2,600 feet MSL while maneuvering to the inbound approach course (7), and a minimum altitude of 1,800 feet MSL (8) while on the inbound approach course prior to reaching the VOR (indicated by the Maltese Cross [9]). The *Minimum Descent Altitude* (MDA) after the VOR is listed, by *Aircraft Approach Category*—A through D—underneath the profile view (10). (See Table 8-1 for a listing of Aircraft Approach categories.) The MDA for all approach categories is 1,340 feet MSL for this approach (11), which is 570 feet AGL (12).

Table 8-1. Aircraft Approach Categories.

Category	Speed
A	90 knots or less
B	91 to 120 knots
C	121 to 140 knots
D	141 to 165 knots
E	166 knots or greater

The numbers after the MDA—1 for Categories A and B, 1 1/2 for Category C, and NA (Not Authorized) for Category D—refer to the minimum in-flight visibility (Part 91) or reported visibility (all other Parts) in Statute Miles required to complete the approach. In the case of the McAlester VOR-A approach, only visibility varies with Approach Category, but that will not always be the case.

Minimum altitudes mean exactly that: you may be higher, but you may not be lower. (An assigned altitude, on the other hand—"Descend and maintain 6000 feet," for instance—implies a tolerance of plus or minus 100 feet.) Once "Cleared for the approach," published minimum altitudes apply, and it is the pilot's responsibility to identify and observe all altitude restrictions—ATC will not assign altitudes or issue descent clearances from that point on. For this reason it is very important that the pilot be able to identify the Minimum Safe Altitude that applies at all points along the approach, from initial maneuvering to final descent and missed approach.

Approach Segments

Every approach can be divided into four major

segments: initial, intermediate, final, and missed. The definitions for each segment are fairly technical (since so many different types of situations have to be included); it is not critical that you understand in every case exactly where, for instance, the initial approach segment ends and the intermediate approach segment begins. It is important, however, that you understand how approaches are organized and that you have a general idea of where each segment begins and ends.

Initial. The initial approach segment is a transition segment from the en route to the approach phase. Since en route and approach situations vary so much, the initial approach segment will vary not only from approach to approach, but will also vary depending upon the direction from which the aircraft approaches the primary nav aid. In the example illustrated, any radial leading To the McAlester VOR would be part of an initial approach segment. Any maneuvering required after that point in order to reverse course and intercept the inbound course (a procedure turn, discussed in detail further on) is also a part of the initial approach segment. Other forms of initial approach segments are DME arcs, outbound radials from nearby VORs, bearings off NDBs, prescribed headings from a nav aid or fix, and radar vectors. The initial segment ends upon interception of the inbound approach course.

Intermediate. The intermediate approach segment is that part of the approach where the aircraft is maneuvering toward the primary nav aid or Final Approach Fix (FAF), either on the inbound course itself, or as part of a specific course designed to intercept the inbound course (a *Terminal Routing*, also discussed in greater detail further on).

The McAlester approach illustrated has no terminal routings—the intermediate segment for the McAlester approach is that portion after the procedure turn from inbound course intercept to the MLC VOR itself. The Minimum Safe Altitude for the intermediate portion is 1,800 feet MSL (8).

Final. The final approach segment is very simple and straightforward: it is that part of the approach from the Final Approach Fix to a point on the airport where either a safe landing can be completed if visual contact has been established, or a missed ap-

proach initiated. The Minimum Safe Altitude for the final approach segment is the MDA for this type of approach: 1,340 feet MSL for all aircraft approach categories.

Missed Approach. All approaches include a missed approach procedure for those times when the pilot is unable to establish visual contact with the airport or runway, and a Missed Approach Point (MAP) is established for every approach in order to positively identify the exact point when the final approach must be abandoned and the missed approach procedure initiated. The missed approach segment is therefore that portion of the approach from the MAP to the holding fix described in the missed approach procedure.

The MAP will be described in different ways depending upon the type of approach. For *non-precision approaches* (no glide slope [GS]), the MAP occurs a specific distance from the FAF. The McAlester approach is a non-precision approach, and the MAP for this approach is 1.5 nautical miles from the FAF (13), a point directly over the runway. If DME information is available in the aircraft, this distance can be measured directly and a missed approach initiated at that point. DME is not, however, required for this approach (or else it would be labeled a VOR DME approach), so an alternate method of identifying the MAP is provided, and that method is dead reckoning; specifically, the pilot estimates the aircraft ground speed and then measures the elapsed time from over the FAF to the MAP. A box listing a range of estimated approach ground speeds and corresponding times is provided (14) for convenience in determining the proper amount of time between the FAF and the MAP.

Some non-precision approaches have the primary approach facility located on the field itself. In these cases, the MAP normally, but not always, occurs with station passage over that facility; the correct MAP will, however, always be described in the MAP box, and it should not be assumed that the MAP automatically occurs over the facility when it is located on the field.

For *precision approaches* (those with glide slopes), the MAP is identified by altitude—the Decision Height CDH); if visual contact has not been established at the Decision Height, a missed approach must be initiated.

The missed approach procedure itself is described just under the profile view (15), and its general direction is indicated by a dashed line and an arrow on the plan view (16). The missed approach procedure will normally be quite simple and direct in order to reduce pilot workload during a critical portion of the approach; its intent is simply to safely extract the aircraft from the missed approach situation and return it to a secure holding area.

The four approach segments create a loop when linked together: the loop begins and ends over the initial approach fix, and extends from the initial approach segment through the intermediate, final, and missed approach segments. A complete approach is therefore a closed system, ensuring a satisfactory alternative regardless of the circumstances (assuming that the approach is flown properly, of course). The loop can be broken only when visual contact has been made with the runway, runway markings, or runway approach lighting (FAR 91.116) and a safe descent can be made to the runway.

The loop can be exited only with an appropriate clearance. Normally this will occur after arrival at the missed approach holding fix; however, in some cases, the controller, in anticipation of a missed approach and in consultation with the pilot, will provide a clearance to an alternate fix or airport prior to or during the approach: "In the event of a miss, fly heading . . ." etc. This expedites and facilitates the missed approach segment and subsequent diversion in the event a miss is necessary.

These are the major elements that all instrument approaches have in common. The rest of the chapter is devoted to those ways in which instrument approaches differ depending on the type of approach aid upon which they are based.

VOR AND NDB APPROACHES

VOR and NDB approaches differ considerably from each other in their practical execution—NDB approaches are less accurate than VOR approaches, are subject to more limitations, and are generally more difficult to track—but they are identical in terms of approach design and general navigational principles. Both are invariably non-precision approaches—electronic vertical guidance is not provided, only Minimum Safe Altitudes—and both involve the interception and tracking of a specific course or courses.

The techniques for intercepting and tracking inbound and outbound courses have been covered in detail in the respective chapters on VOR and NDB navigation, and will not be repeated here. Suffice it to say that these principles and techniques must be thoroughly understood, or it will be very difficult to fly an accurate VOR or NDB approach.

Non-Radar Environments

In a non-radar environment, the initial approach fix for a VOR or NDB approach is often the primary approach facility itself (MLC VOR in our previous example), although it can also be a nearby VOR, NDB, or en route intersection. It is from over this fix that the actual approach begins. From over the initial fix the aircraft normally must be maneuvered into a position where the inbound approach course can be intercepted. There are two ways to accomplish this positioning—through a terminal routing, or with a procedure turn (PT).

An example of a type of terminal routing is shown in Fig. 8-2, a second approach to McAlester, this one a VOR DME approach to runway 19. (This approach is similar in many ways to the McAlester VOR-A approach, except: DME is required for this approach; it is conducted from the north rather than the south; this approach leads to a specific runway, not just to the airport.) At the top of the procedure are two DME arcs with the note "17 DME Arc" written over each arc and "3000 NoPT" written underneath. This note means that an arc based on a distance of 17.0 DME from the MLC VOR, at a minimum altitude of 3,000 feet MSL, can be used to intercept the inbound approach course directly, without the necessity for a procedure turn. (In fact, a procedure turn will not even be allowed, unless specifically requested and authorized.) Other forms of terminal routings are radials off nearby VORs, bearings from nearby NDBs, and headings published as a part of the approach procedure. Whatever the form, all terminal routings can be identified by the note

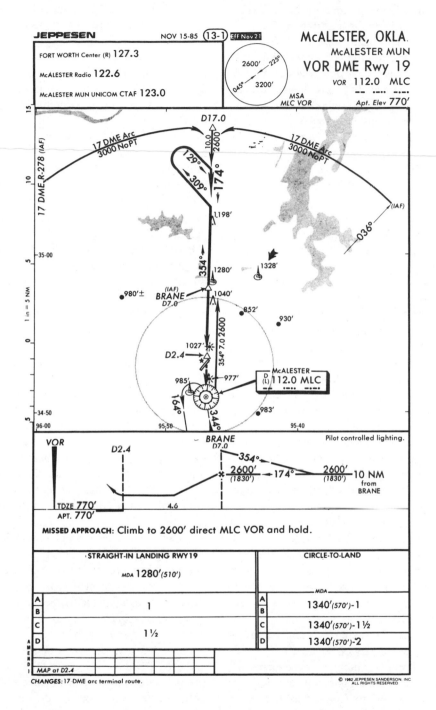

Fig. 8-2. When a particular routing is marked "NoPT", the routing is called a "terminal routing." Inbound approach courses can be intercepted directly from terminal routings. Reproduced with permission of Jeppesen Sanderson, Inc. Not actual size. Not to be used for navigation.

"NoPT." Terminal routings are convenient and direct, and generally provide for an easy transition from the en route to the approach phase of flight.

Where a terminal routing is not available, or is too far to the other side of the approach for a given direction of flight to be useful, a procedure turn can be used instead to intercept the inbound course from a starting point over the initial approach fix. A procedure turn is a course reversal away from the primary fix; its purpose is to maneuver the aircraft into a position where the inbound course can be intercepted.

There are only two restrictions on procedure turns: all maneuvering must be done on the side indicated on the plan view (the west side for both McAlester approaches), and all maneuvering must be done within the prescribed limits for the procedure (10 nautical miles in both cases). As long as these two restrictions are observed, any kind of turn can be used.

The most common way to execute a procedure turn is to fly outbound from the initial fix on the inbound course, then turn to a heading that is 45 degrees away from that course, followed by a 180-degree course reversal. At the completion of the course reversal the aircraft will be on a 45 degree angle to the inbound course, which is normally an ideal interception angle. The 45/180/45 degree procedure turn is, in fact, so common and so effective, that both Jeppesen and NOS approach charts provide the necessary headings for this type of procedure turn on the approach plate itself—309 and 129 for the McAlester VOR DME approach.

There are other ways to do procedure turns though. Common alternatives are the teardrop pattern, the 80/260 degree procedure turn, and the racetrack pattern. These alternate procedure turn types are illustrated in Fig. 8-3. (When an alternate procedure turn pattern is specifically depicted on the approach chart, then it must be observed.)

The teardrop pattern is simpler than the conventional 45 degree pattern—fly outbound 30 degrees or so off the inbound course, then reverse course—but course intercept is less predictable. It is only designed into an approach when, for some reason, a large amount of altitude must be lost after the initial approach fix, or when another facility, such as a VOR,

exists next to the FAF, making it convenient. As an option to the standard pattern, its main virtue is simplicity.

The 80/260 pattern consists of an initial turn to a heading 80 degrees off the inbound course, followed by an immediate 260 degree turn back around again. It is the fastest type of procedure turn, and the easiest to get into trouble with. It works best when a strong tailwind component exists outbound, since it tends to keep the turn close to the primary fix and minimizes the tendency to extend too far out. With a strong headwind component outbound, position relative to the primary fix must be monitored carefully, or the aircraft may remain too close to the fix, perhaps even drifting inside it to the final approach side of the fix. With a strong crosswind component from the procedure turn side (i.e., a headwind on the 80 degree part, and a tailwind on the 260 degree part), the pilot must be very careful not to miss the inbound course entirely: in these conditions, lateral movement away from the inbound course is minimal, while re-interception speed is maximized by the tailwind—a double negative situation sure to catch the pilot who is not prepared for it.

The racetrack pattern is convenient when a direct entry into the pattern can be made; i.e., when approaching from more or less directly behind the FAF. In these cases, the purpose of the procedure turn is not to reverse course, but to lose altitude, and the racetrack pattern is a simple and effective way to lose 2,000 to 3,000 feet of altitude. If more altitude than that needs to be lost, the pattern can be repeated.

Experienced instrument pilots take advantage of the strengths inherent in each type of pattern to facilitate course intercept as conditions dictate, but, in general, these alternate types of procedure turns should not be attempted by inexperienced instrument pilots unless required as a part of the approach procedure. While they have their advantages in certain situations, they also have their disadvantages in others, and unless you know exactly what you are doing, it is much safer and simpler to use the standard 45 degree procedure turn pattern depicted on the charts.

Procedure turn inbound is that portion of the approach where course reversal has been completed

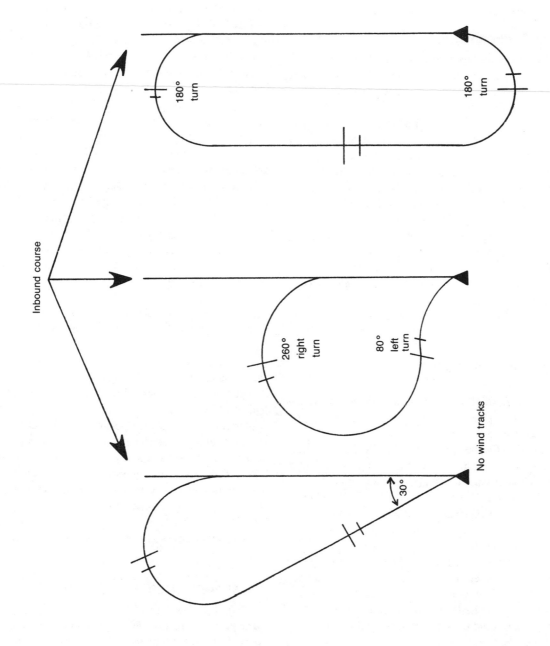

Fig. 8-3. Alternate types of procedure turns. From left to right: the teardrop pattern, the 80/260, and the racetrack pattern.

and the inbound approach course has been intercepted. Procedure turn inbound is a common reporting point for aircraft conducting full approaches in non-radar environments, since it marks the beginning of the intermediate approach segment.

Radar Environments

In a radar environment, vectors to intercept the inbound approach course will normally be provided by the radar controller. The controller will assign headings to fly in order to maneuver the aircraft into position for the inbound course intercept. The final interception angle will normally be between 30 and 45 degrees, and will be gauged to intercept the inbound course just outside of the Final Approach Fix (unless the controller advises otherwise or you have requested either a close-in or an extended turn-on). Radar vectors simplify the initial and intermediate segments of approaches enormously, and expedite traffic at busy airports, but the primary responsibility for navigation still rests with the pilot, and the pilot must be prepared to resume non-radar procedures at any point if it becomes necessary (i.e., proceed directly to the VOR and conduct a full, procedure turn type approach.)

Inbound, Final Approach, and Missed Approach

Regardless of how the aircraft arrives at the inbound course—radar vectors, a terminal routing, or a procedure turn—once established on that course (and cleared for the approach), the pilot is free to descend to the minimum altitude listed for the inbound approach segment. Upon arrival over the Final Approach Fix (the VOR in this case), the pilot should report his position (in a non-radar environment), begin timing, and descend to the Minimum Descent Altitude. Upon reaching the MDA the pilot should begin to look for the airport, or, for a straight-in type of approach to a specific runway, the runway, runway markings, or approach lights. If visual contact has not been established upon reaching the Missed Approach Point (and even if visual contact has been established, but the aircraft is not in a position from which a normal descent and landing can be made), then the pilot is obligated to initiate a missed approach. The missed approach will normally be a return to a published hold over the primary approach facility.

The missed approach point and the first part of the missed approach procedure should be memorized. The missed approach is a critical point, since the aircraft is slowest, dirtiest, and closest to the ground at that point; searching for chart information at that point is distracting at best, and it can lead to vertigo and disorientation, with little chance of recovery.

There are many variations on this basic format for a non-precision VOR or NDB approach: sometimes, when DME is not required but is available, a lower MDA may be authorized. *Step-down* fixes are sometimes provided: an intermediate minimum altitude authorized after passage of an NDB bearing, a fan marker, a DME fix, a radar fix, or a radial from a nearby VOR. In some cases, the primary facility will be located directly on the field; in this case, the Final Approach Fix becomes a Final Approach Point (FAP). The FAP is not a specific fix, but is simply that point where the aircraft is established inbound in position to begin its descent to the MDA. These variations are necessitated by the many geographic and physical variables inherent in any approach situation.

Pilots planning to file instrument flight plans should study the applicable approach charts for their destination and alternate airports as a normal part of their preflight planning routine. Most questions about approaches can be answered by a careful reading of the introductory material supplied with the approach charts, and, in particular, by the legend descriptions and definitions. Pilots having additional questions about a particular aspect of an approach should consult with an experienced instrument pilot, preferably either a Certified Flight Instructor, Instrument or an Airline Transport Pilot, prior to filing the flight plan.

ILS APPROACHES

The ILS (Instrument Landing System) approach is the primary precision approach facility for civil aviation. The initial segment of the ILS approach is very similar to the initial segment for VOR and NDB approaches (i.e., interception of the inbound course via radar vectors, terminal routings, or procedure turns); however, the intermediate and final approach segments are quite different from VOR and NDB intermediate and final approach segments, since an ILS is itself quite different from a VOR or an NDB.

On an ILS, navigational information is provided in the form of marker beacons or DME for range, a localizer course for azimuth, and a glide slope for elevation. Approach lighting systems of various configurations are installed to facilitate the transition from instrument to visual conditions. Compass Locaters—low powered Non-Directional Beacons—are frequently located on the localizer course to facilitate orientation while maneuvering for the approach, and to provide the pilot with a rough indication of

the inbound course intercept point. (Compass Locaters are not, however, required ILS components.) Standard characteristics and terminology for an FAA ILS are illustrated in Fig. 8-4. An example of an actual ILS approach chart is illustrated in Fig. 8-5.

ILS Minimum Altitudes

Since the ILS is a precision approach, a minimum altitude for the intermediate segment does not apply, nor does a Minimum Descent Altitude apply to the final approach segment. Instead, the last assigned or procedure turn altitude is maintained until the glide slope has been intercepted. (The glide slope is displayed to the pilot by either a horizontal needle

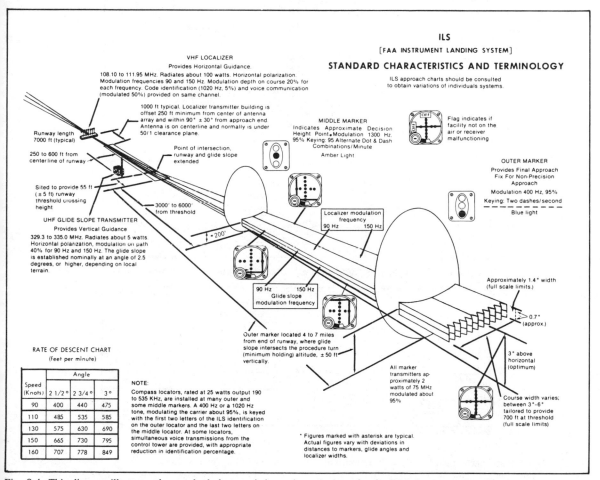

Fig. 8-4. This diagram illustrates the standard characteristics and terminology for the FAA Instrument Landing System.

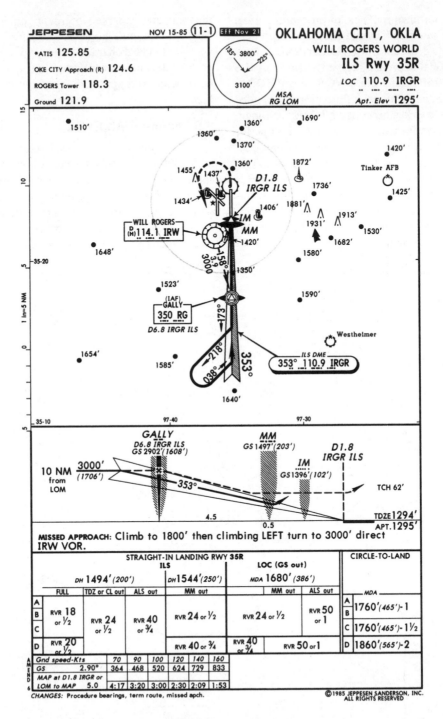

Fig. 8-5. This ILS approach, a particularly well equipped example, incorporates a Compass Locator, DME, and an Inner Marker. These are not required components, but they can be helpful in orientation and position awareness when they are available. Reproduced with permission of Jeppesen Sanderson, Inc. Not actual size. Not to be used for navigation.

on the VOR CDI, shown in Fig. 8-6, or by vertical pointers on the FD/HSI, shown in Figs. 8-7 and 8-8.) Once intercepted, the glide slope needle or pointer is tracked vertically in much the same manner that the VOR needle is tracked horizontally: pitch is adjusted in the direction of the needle off center (within performance limits and supplemented by power).

Descent is an automatic function of glide slope tracking, and intermediate descent altitudes are not necessary. In place of an MDA, a Decision Height (DH) is established for each ILS approach, and the descent is continued along the glide slope until Decision Height has been reached. (The DH for the ILS approach illustrated in Fig. 8-5 is 1494 feet MSL, which is 200 feet AGL.) At the DH, if visual contact with the runway or the approach lighting system has been established and the aircraft is in a position where a safe landing can be accomplished, the approach can be continued visually and the landing completed; otherwise, a missed approach is executed.

The actual DH will vary with each approach and is primarily a function of terrain and obstructions. The lowest DH authorized for a standard, CAT I ILS approach is 200 feet above the terrain. Lower DHs are authorized for CAT II and CAT III approaches,

Fig. 8-7. Flight Directors typically display glide slope with a pointer to one side or the other of the main display—here on the left side. In the approach mode, any deviation from the approach course or glide slope will be reflected on the FD command bars. Photo courtesy of Collins Avionics, Rockwell International.

down to a Decision Height of zero, however these approaches require special authorization and equipment. (CAT III approaches are further broken down into CAT IIIA, B, and C, with the CAT IIIC approach having no visibility minima, and no Decision Height—a "zero-zero" approach.)

ILS Azimuth

Azimuth, provided by the localizer component of the ILS, is tracked in a manner similar to a VOR course, with two important differences: one, the localizer course consists of a single course only, and two, the localizer course is four times as sensitive as a VOR course. In practice, the same principles apply to tracking a localizer as do to tracking a VOR, but greater care is required, and greater accuracy is provided.

ILS Range Information

Range information for an ILS is normally provided in the form of marker beacons, specifically, an Outer, a Middle, and sometimes an Inner marker. (DME is co-located with some ILSs, including this

Fig. 8-6. A complete VHF navigation and communication package for general aviation. The horizontal needle on the VOR Cockpit Deviation Indicator at the top left is a glide slope deviation indicator. Photo courtesy of Narco Avionics.

Fig. 8-8. Horizontal Situation Indicators typically display glide slope data with a pointer to one side of the display. The control panel below can be used to control course pointers and DME display for two different HSIs, as well as movement of the heading "bug" on both HSIs. Photo courtesy of Collins Avionics, Rockwell International.

one, but these tend to be the exceptions.) Marker beacons are low powered beacons radiating a signal in an elliptical pattern. When a suitably equipped aircraft passes over a marker beacon, a corresponding light is illuminated in the cockpit, and a coded tone is emitted through the cockpit speaker or headphones. The outer marker transmits a 400 Hz tone in a pattern of continuous dashes, and illuminates a blue light. Its main purpose is to indicate that point where the glide slope intercepts the procedure turn altitude. The outer marker also serves as the FAF when the glide slope is inoperative or unavailable.

The middle marker transmits a 1300 Hz tone (i.e., higher in pitch than the outer marker) in a pattern of dots and dashes, and illuminates an amber light. The middle marker is placed approximately at the point where the DH would normally be reached for an aircraft exactly on the glide slope. It thus serves as a backup to and confirmation of the Decision Height. It is not, however, a substitute for the Decision Height: decision height is strictly a function of altitude and a missed approach must be initiated at

DH if visual contact has not yet been made, even if the middle marker has not been reached.

Inner markers—3000 Hz, all dots, at a rate of six per second, illuminating a white light—are used to mark Decision Heights on CAT II and CAT III approaches. On a standard ILS (CAT I), if an inner marker has been installed, it simply indicates that the aircraft is somewhere between the middle marker and the runway threshold (normally a point over the middle of the approach lighting system).

LOCALIZER APPROACHES

Whenever glide slope is not available, either because of equipment failure on the ground, glide slope receiver failure in the aircraft, or lack of a glide slope receiver/indicator in the aircraft (glide slope is not required for instrument flight), the ILS automatically becomes a *Localizer* approach. In some cases, otherwise complete ILSs are commissioned without glide slope by design; these are called Localizer approaches, and are abbreviated LOC on the approach charts. For whatever reason—by design or by failure—an ILS without glide slope is a Localizer approach.

Localizer approaches are non-precision approaches. They are very similar to VOR approaches, and are flown in a similar manner, except that a localizer course is four times as sensitive as a VOR course, and LOC approaches normally include marker range indicators, while VOR approaches normally do not. Minimum intermediate altitudes apply, and final descent is done to a Minimum Descent Altitude. Localizer minimum altitudes are always published on the ILS approach chart on the right hand side of the minima box, under "LOC (GS out)", so that the pilot can use the ILS as a LOC approach when glide slope is not available, and can continue the approach as a LOC approach should the glide slope fail during the approach.

Localizer approaches fill the gap between basic, non-precision approaches such as VOR and NDB approaches, and precision approaches such as ILS approaches, and share characteristics of each.

LDA AND SDF APPROACHES

Variations on the Localizer approach are the

LDA—the Localizer-type Directional Aid—and the SDF—a Simplified Directional Facility. An LDA is a localizer that is not aligned with a specific runway—the inbound course is offset somewhat from the runway, the exact amount of the offset varying from slightly to considerably. An SDF is very similar to an LDA except that course sensitivity is reduced by about half—something between a localizer and a VOR course in sensitivity. Neither of these two LOCALIZER type variations is particularly common, but they are encountered occasionally and are something the instrument rated pilot should be familiar with. LDAs and SDFs fill the small gap between VOR and Localizer approaches.

CIRCLING APPROACHES

Whenever possible, approach courses are aligned to coincide with a specific runway; however, due to physical and economic restraints, this is not always possible. Whenever the final approach course and the landing runway differ by more than 30 degrees, the approach is considered to be a circling, rather than a straight-in approach. The VOR-A approach to McAlester, for instance, is a circling approach. (Circling does not mean that the aircraft must be literally flown in a circle, only that maneuvering in excess of a 30 degree course change is required.)

The ILS approach illustrated in Fig. 8-5 is a straight-in ILS to runway 35 Right; it can also be used as a circling approach to runways 35 Left (the parallel runway), runways 17 Left and Right (the other ends of these parallel runways), and runway 12 and 30 (the shorter runway that intersects 35 Left/17 Right). Circling minimums are inevitably higher than straight-in minimums, and a separate section for those minimums is provided to the far right of the minima section, specifically labeled "Circle-To-Land." Circling minimums are higher than straight-in minimums because the circle-to-land maneuver is a visual procedure done largely without the aid of electronics. Both higher ceilings and greater visibility are required for it to be safely accomplished.

Since the size of the maneuvering area is a direct function of airspeed (the greater the airspeed, the wider the turn radius), circling minimums are further broken down into approach categories based on

airspeed. Approach categories are labeled A through D (the military has a category E as well), and as can be seen from the circling minimums boxes, the minimums increase in altitude and/or visibility with each increase in category.

Table 8-1 is taken from the *AIM*; it defines and lists the airspeed brackets for each category, based on 1.3 times the stall speed in the landing configuration (V_{so}) at maximum gross landing weight. However, approach category is based on the circling airspeed actually flown, not the minimum theoretical airspeed. Thus a single engine aircraft with a V_{so} of 60 knots and a calculated circling airspeed of 78 knots would normally be a Category A aircraft, but if it is actually flown at 100 knots while circling to land, then the pilot would have to use Category B minimums.

Pilots can elect to use higher category minimums, even if not required; some operations, as a matter of policy, use Category D minimums regardless of aircraft type and even if actually circled at a slower airspeed. Using a higher category than is legally required sometimes results in a diversion to an alternate that might not otherwise have been necessary, but it adds an extra margin of safety to what can be a tricky procedure, particularly if the pilot is unfamiliar with the airport.

MICROWAVE LANDING SYSTEM

The next generation of precision approach aids is called the Microwave Landing System, or MLS. In many ways, an MLS is simply a very much improved ILS, but it does have certain capabilities that the ILS does not have, and it is therefore a different type of precision approach from an ILS. We will look first at the way in which MLS improves upon ILS, and then at the ways in which it is different from ILS.

MLS Improvements

MLS has improved upon ILS in nearly every respect. With MLS, azimuth angle—the equivalent of the localizer course for an ILS—can vary by as much as 60 degrees to either side of the associated runway. Lateral guidance—right/left indication—is provided across the entire course width (as opposed to 10 degrees to either side for an ILS). This provides great flexibility around terrain and obstacles on the straight-in approach path and allows for multiple approach courses—aircraft can approach, for instance, in alternating fashion from either side of the extended runway centerline.

MLS vertical guidance is called glide path (rather than glide slope, for some reason known only to the FAA—actually, there is a minor distinction, but the reason for it is still somewhat of a mystery to me). MLS glide path can be varied from 0.1 to 15.0 degrees: helicopters can descend on a 15-degree glide path, for instance, STOL aircraft on a 10-degree glide path, piston aircraft on a 3.5- or 4.0-degree glide path, and turbine aircraft on a 2.5- or 3.0-degree glide path. Every aircraft type can use the glide path angle that is most appropriate for its own performance characteristics.

Range along the MLS course is provided not by markers, but by a precision DME system (DME/P) that is accurate to within 100 feet. (It is anticipated that the requirement for DME/P will be waived at a few outlying, low density MLS installations, however, precision range information will normally be an integral part of the MLS system.) With DME/P, the pilot will have extremely accurate distance-to-go information at all times while on the MLS approach—a great improvement over range markers.

Physically, MLS ground facilities are much simpler to install than ILS facilities are, are not critical in terms of terrain, and are virtually immune to interference from vehicles on the ground (not the case for ILS, where a truck passing in front of the localizer antenna, for instance, can cause a momentary needle deflection in the cockpit). The MLS glide path antenna does not require a snow-free ground plane as the ILS glide slope does, and MLS is relatively free of weather-induced errors.

Finally, the MLS system will have 200 channels, versus 40 for the ILS system, which will allow for virtually an unlimited number of MLS installations within the same area of signal coverage without interference. High density areas, such as New York and Los Angeles, are currently running out of free ILS frequencies. This will not be the case with MLS.

One of the first MLSs installed was at Cadillac, Michigan. That approach plate is illustrated in Fig. 8-9. An MLS approach plate looks very much like an ILS approach without markers. Markers are replaced by DME/P, and MLS frequency selection is by channel (Channel 602 here). All MLS identification codes begin with *M* (dash dash), just as all ILS identification codes begin with *I*.

ILS is a very effective precision landing system, and has served aviation well over the years, but it does have its limitations; MLS is designed to eliminate those limitations. It is expected that about 1,200 MLSs will be fully operational by 1995. This compares to about 1,000 ILSs currently in use.

MLS Differences

Since MLS has the capability to vary azimuth and elevation, MLS operates a little differently from the pilot's point of view than does ILS. With ILS, a single frequency is tuned, and a single localizer course and glide slope angle is automatically provided—the only one available. With MLS, the pilot can select different approach courses and glide path angles (within approach limits); therefore, the MLS receiver includes not only channel selection for automatic tuning of the standard azimuth, elevation, and DME/P signals, but also has provision for entry of optional course azimuth variation and glide path angle. A typical MLS control/display unit is illustrated in Fig. 8-10; the top window is for channel selection (CH 516 in this example), the middle window is for glide path angle (6.6 degrees), and the bottom window is for course (340 degrees).

MLS has an additional approach capability that ILS does not have, and that is the capability to fly multiple and even curved approach paths. Fig. 8-11 shows what a typical multiple approach path might look like. In this hypothetical example, a course of 160 degrees is flown from over Indio, an Initial Approach Fix, until reaching Tiney intersection. Then, on a prescribed heading of 135 degrees, the straight-in approach course of 180 degrees for Runway 18 is intercepted, and the approach completed.

Segmented approaches may prove to be very advantageous in mountainous terrain, since this will allow approach paths to be designed that follow the low lying areas and avoid the peaks and ridges. Segmented approaches should also provide greater flex-

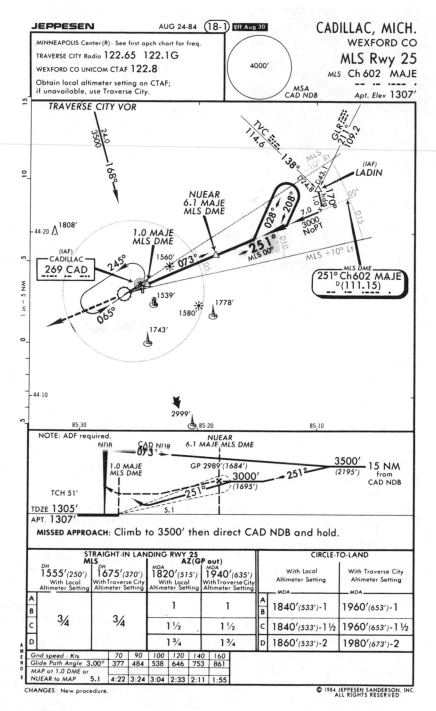

Fig. 8-9. An MLS approach plate is very similar to an ILS approach plate, however range is indicated by precision DME rather than by marker beacons, and frequency selection by channel (Channel 602 in this case). MLS identifiers start with the letter ''M.'' Reproduced with permission of Jeppesen Sanderson, Inc. Not actual size. Not to be used for navigation.

Fig. 8-10. This Sperry MLS Control/Display Unit illustrates the unique features of the Microwave Landing System. Channel selection is entered in the top window. Glide slope and course selections are entered in the next two windows. Precision DME is automatically tuned. Copyright Sperry Corporation, 1985.

ibility in air traffic control, which will result in expedited arrivals and departures, and will help with noise abatement and the avoidance of restricted and prohibited areas.

It is even possible to create curved approaches with MLS by integrating multiple approach paths with an onboard computer to continuously resolve and display to the pilot a constantly changing inbound course. The pilot will "fly the needles," as always, but the aircraft will actually track a curving course to the airport. Curved approaches offer even greater flexibility in approach planning than segmented approaches do.

Finally, MLS has the capability for basic and auxiliary data transfer from ground to aircraft. Basic data transfer will be available at all MLS installations and will include station identification, exact transmitter/antenna locations (information essential for MLS functioning), performance level and status, and DME/P channel and status. Auxiliary data, available at high-density MLS installations, will include three-dimensional position, waypoint coordinates, runway conditions, and current weather.

In short, MLS promises to be everything ILS al-ready is, but it will be more accurate, more reliable, more stable, easier to install and maintain, more flexible in approach design, freer of environmental and physical limitations, and will offer expanded data transfer capability.

LORAN-C APPROACHES

LORAN-C accuracy is a direct function of aircraft location in relation to Master and Slave stations. (See Chapter 9 for a complete discussion of LORAN-C navigation.) At certain locations, LORAN-C accuracy is sufficient, at least in theory, to provide nonprecision approach guidance at least as good as that provided by VOR and NDB, and in many cases better. This opens up tremendous possibilities for improvement in non-precision approach guidance for many locations. Many airports lack approach guidance altogether, and others have only limited VOR or NDB approach capability with fairly high minimums. These airports may be able to take advantage of favorable LORAN-C information for greatly improved instrument approach capability. Not all airports will qualify (those on the Plains states, where LORAN-C coverage is not presently available, are obvious examples), and, in any case, this is still an experimental concept with limited testing and experience at the time of this writing, but it is very possible that LORAN-C approaches will be as common at some point in the future as VOR approaches are now.

SATELLITE APPROACHES

While LORAN-C offers much promise as an approach aid, satellite navigation—NAVSTAR/GPS—offers perhaps even more. (Satellite navigation is covered in Chapter 12.) NAVSTAR/GPS will be a three-dimensional system, and this capability will provide for the possibility of a precision NAVSTAR/GPS approach.

NAVSTAR/GPS is primarily a military system, and the only mode available to civil aviation (the C/A mode) will not be able to provide vertical accuracy equal to that provided by either ILS or MLS; therefore, precision NAVSTAR/GPS approaches for civilian aviation will not be as accurate as existing

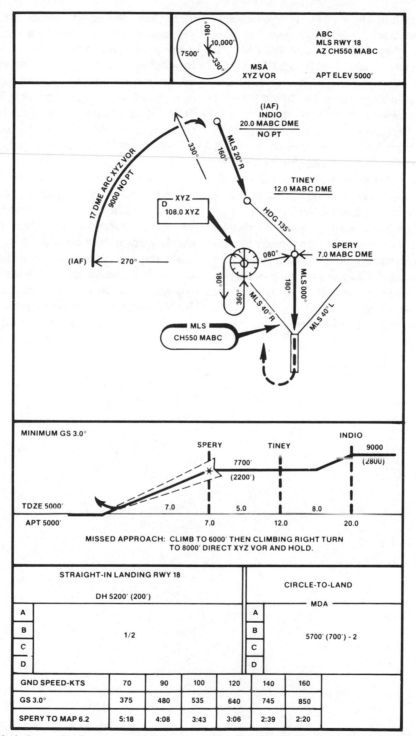

Fig. 8-11. This hypothetical example of a segmented MLS approach illustrates the multiple approach path capability inherent in the Microwave Landing System. Copyright Sperry Corporation, 1985.

precision approaches. The availability of even semi-precision approaches at low density locations currently served only by VOR and NDB still will be a great improvement. In any case, NAVSTAR/GPS's global coverage (when fully operational) will make at least non-precision approach capability available for any airport, anywhere in the world.

If satellite navigation is somehow improved to the point where precision approaches equal to ILS and MLS standards can be created (for instance, by opening up the precision, or military mode to civilian use, or perhaps by the creation of a separate, civilian satellite system), then the very concept of an instrument approach as a separate procedure, distinct from en route navigation, may disappear: the approach phase will merely be the final segment of a continuous, satellite-based navigational process that begins and ends on the ramp, with no ''seams'', or transitions, in between.

Much work remains to be done in this area—testing for satellite approaches has not even begun yet—and it will no doubt be several, perhaps even many years before satellite approaches are a reality. In addition, as with any new system, problems may arise that may diminish or even negate its utility as an approach aid. Still, it is possible that at some point in the future there will be only one type of approach, a precision NAVSTAR/GPS approach. In any case it is very possible that at some point in the future no more than two types of approaches will exist: precision MLS approaches, and non-precision NAVSTAR/GPS approaches. Since most aircraft accidents occur during the approach and landing phase of flight, any change in approach capabilities that simplifies and standardizes instrument procedures will almost certainly translate into an an immediate and direct increase in aviation safety. With precision MLS approaches at many locations, and non-precision NAVSTAR/GPS approaches universally available, the improvement could be dramatic.

Chapter 9

LORAN-C Navigation

LORAN-C (long range navigation, Version C), is revolutionizing navigation for general aviation. What INS (inertial navigation—see Chapter 11) did for the airlines and OMEGA/VLF (Chapter 10) did for corporate aviation, LORAN-C is now doing for private general aviation: providing a practical system of long-range area navigation at a cost that is appropriate to the level of operation. LORAN-C is not perfect—there are still some serious gaps in coverage in the United States and LORAN-C falls short of being a truly worldwide system but these shortcomings are recognized and are being addressed.

Ten years ago virtually no one in aviation had even heard of LORAN. LORAN systems did exist (LORAN-A), but they were limited in range, were subject to many errors, had many limitations (LORAN-A operated in the Medium Frequency band just above the AM broadcast band), were based on pre-computer chip technology, and were intended primarily as marine navigational systems. LORAN-C changed all that. Although LORAN-C is based on the same general principles as LORAN-A, it is a much more sophisticated system than LORAN-A.

LORAN-C operates in the relatively stable Low Frequency band, and it is able to provide very long range coverage (up to 2,800 nautical miles over water) with many fewer limitations. Perhaps most importantly, LORAN-C takes advantage of computer chip technology to create a system that is fast, practical, and easy to use. (LORAN-A systems were generally quite slow and were fairly difficult to use, requiring interpretation and plotting of data obtained from the LORAN-A receiver—suitable for slow moving vessels, but not suitable for non-navigator crewed aircraft.) LORAN-C is still primarily a marine system and is still under the operational control of the US Coast Guard, but LORAN-C is suitable for aircraft use, where LORAN-A generally was not.

As an aside, there is a third type of LORAN, called LORAN-D. It is very accurate (plus or minus 600 feet versus a plus or minus maximum of 2.5 miles for LORAN-C), but is limited in range. It is a military system, designed to provide position information in tactical situations: close air support and interdiction, reconnaissance, air drop, and rescue. LORAN-D currently has no application to civilian air navigation, and, other than this brief mention, is not covered in this book. LORAN-A has been discontinued. (If a LORAN-B system ever existed, I am unaware of it; none exists at present.)

LONG RANGE NAVIGATION

LORAN-C is a long range system of area navigation. That does not mean that it cannot be used over short ranges, nor does it even mean that it works best over long ranges, but it does mean that LORAN-C can be used over long ranges, where shorter range systems like VOR cannot. There are fundamental differences between long and short range navigational systems; this next section will look at some of those differences before moving on to the specifics of LORAN-C theory and operation.

Great Circle Routes

The most important difference between a short and a long range navigational system is that a short range system can treat the earth as flat, disregarding the small errors introduced in so doing for simplicity's sake, but a long range system cannot—the errors over long distances become too significant to disregard. Over short distances, for instance, Rhumb Line courses are perfectly adequate; over long distances, Great Circle courses are required if substantial errors in routing and distance are to be avoided. (Refer back to Chapter 1, Pilotage and Dead Reckoning, for a review of the difference between a Rhumb Line and a Great Circle course.)

The difficulty with flying Great Circle routes is that the course bearing changes with each line of longitude crossed. LORAN-C systems automatically plot and keep track of bearing changes with changes in longitude, so this problem is resolved for the pilot. With LORAN-C, the pilot can fly a Great Circle route as easily as he can track a VOR radial.

Magnetic Variation

Changes in Magnetic Variation can be substantial over long distances. (VOR compensates for magnetic variation by aligning each facility with Magnetic North. No further correction on the part of the pilot is required.) Most LORAN-C units automatically keep track of and correct for magnetic variation. Some provide automatic correction only within the Continental United States; others require manual entry of magnetic variation for all locations by the pilot. Whatever the system capability, changes in magnetic variation must be accounted for and resolved with long range systems—magnetic variation changes, for instance, from 20 West in the northeast corner of Maine to 21 East in the northwest corner of Washington—a difference of 41 degrees.

Latitude/Longitude

A further difference between long and short range systems is that, as a purely practical matter, long range systems need a universal system to describe fixes—names, radials and distances work satisfactorily in identifying and describing short range fixes, but to try to find a fix that could be anywhere on the surface of the earth by name, radial, and distance is unsatisfactory.

The best system for describing fixes is the latitude/longitude system of global coordinates. Most pilots are somewhat familiar with the latitude/longitude ("lat/long") system of describing position on the surface of the earth, but it is important that they understand this system thoroughly if they intend to use a long range system of navigation.

The earth rotates about its North/South axis; the plane that is perpendicular to the midpoint of that axis is the Equator. (See Fig. 9-1.) The North/South axis and the Equator are the two basic references used to describe the location of any point on the surface of the earth. The Equator logically divides the earth into two horizontal halves, a Northern and a Southern Hemisphere. The division of the earth into vertical halves is not so logical, and the point of division has to be made arbitrarily. For historical reasons, the division that has been decided upon is a Great Circle (called a meridian) whose plane passes vertically through a point where an early observatory was located: Greenwich, England. This meridian represents the "vertical equator." The half of that plane that extends through Greenwich is the zero meridian (commonly called the Prime Meridian). The other side of that meridian—the extension that runs vertically through the Pacific Ocean—is the 180 degree

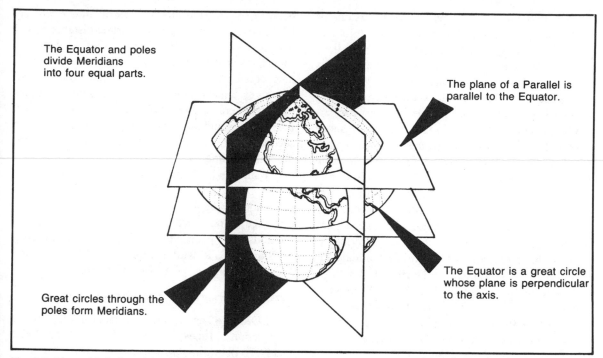

Fig. 9-1. Lines of latitude and longitude are based on planes formed by the Equator and the zero meridian.

In the figure:

The Equator and poles divide Meridians into four equal parts.

The plane of a Parallel is parallel to the Equator.

Great circles through the poles form Meridians.

The Equator is a great circle whose plane is perpendicular to the axis.

meridian, also known as the International Date Line. (The International Date Line is the point where the beginning of one day catches up with the end of the previous day. A globe, and very strong powers of concentration, are normally required to see how the International Date Line works. It is all right to take it on faith.)

These two planes divide the earth into four quadrants: North/East, North/West, South/East, and South/West. The Continental United States lies entirely within the North/West quadrant: North of the Equator; West of the zero meridian.

To further identify any given point, lines running parallel to the Equator, called Parallels of Latitude, or more commonly simply "Parallels," describe location in relation to the Equator. The Equator is the zero parallel, and there are a total of 90 degrees of latitude between the Equator and each of the poles. Thus parallels of latitude consist of a number from 0 to 90, 0 being the equator and 90 either the North or South Pole. In order to determine on which side of the Equator a parallel lies, a North/South identifier is included with the number

of degrees: 45 degrees North is a parallel halfway between the Equator and the North pole; 45 degrees South is a parallel halfway between the Equator and the South Pole. Parallels are evenly spaced between the Equator and each of the poles, and are therefore an equal distance apart; in fact, there are exactly 60 nautical miles between each degree of latitude, and a single minute of latitude (1/60th of a degree) is one nautical mile.

The vertical equivalent of parallels of latitude are meridians of longitude. Starting from the zero meridian and moving east or west, meridians are labeled from zero to 180: 10 degrees West is a line running through Ireland and Africa; the 140 degree East meridian runs through Japan and Australia.

Meridians are not drawn parallel to the zero meridian, but radiate instead from each of the poles, like ribs from an umbrella. Meridians are therefore closest together at the poles, and widest apart at the equator. Accordingly, a minute of longitude is therefore not a constant distance: it is equal to one nautical mile only at the equator, diminishing in size with distance from the equator. Meridians are said to converge as they approach the poles, and this is one of the reasons polar navigation is difficult—a pilot flying close to the North Pole crosses many meridians of longitude in a very short period of time (just as a person walking around the North Pole can go around the world in a few steps.) It might be easier at times if meridians were parallel, at least for certain purposes, and there is in fact a fairly advanced system of navigation sometimes used in polar regions called Grid Navigation which uses North/South parallels instead of meridians. If you would like to know more about this system, see the Air Force/Navy Air Navigation training manual listed in Appendix A.

With this system of parallels and meridians, any point on the surface of the earth can be described in terms of the intersection of a parallel of latitude and a meridian of longitude: 45 North 30 West is a specific point in the middle of the North Atlantic; 15 South 75 East is a specific point somewhere in the Middle of the Indian Ocean. (By convention, the order of coordinates is always latitude first and longitude second—alphabetical order.)

Degrees of latitude and longitude are also broken down into 60 minutes (′) per degree, and 60 seconds (″) per minute; seconds, however, are not commonly used in air navigation—tenths and sometimes hundredths of a minute are used instead. (A tenth of a minute is 607.6 feet. This is accurate enough to differentiate among different areas on an airport, such as between the ramp area and the runway area; a hundredth of a minute is 60.8 feet, or accurate enough to locate the touchdown zone of a specific runway.) The Kennedy VOR (JFK, 115.9 MHz)) is located on the 40 degree, 38 minute 00 second parallel North (N 40 38′ 00″), and the 73 degree, 46 minute 24 second meridian West (W 73 46′ 24″). In air navigation, this would normally be shortened to "North 40 38.0, West 073 46.4," or simply: N40 38.0 W073 46.4. (The zero is added before 73 because longitudinal coordinates range from 000 to 180, while latitude varies only from 00 to 90.)

To summarize: parallels of latitude run from zero at the Equator to 90 North or South; meridians, or longitudinal lines, run from zero at the zero meridian (which runs through Greenwich, England) to 180 East and West (which is on the other side of the globe opposite the zero meridian). Lines of latitude run parallel to each other and are a constant distance apart: one degree equals 60 nautical miles in distance. Meridians converge at the poles: a degree of longitude equals 60 nautical miles at the Equator, but decreases with distance from the equator. Degrees of both latitude and longitude are further broken down into minutes and seconds of 60 units each, however seconds are frequently converted into tenths or hundredths of minutes.

With this background information in mind, the next two sections examine the theory of LORAN-C operation and show the existing areas of LORAN-C coverage. The remainder of the chapter covers methods and specific techniques of operation, both VFR and IFR, as well as LORAN-C limitations and accuracy, and prospects for the future.

LORAN-C THEORY OF OPERATION

LORAN-C determines present position from the intersection of Lines of Position (LOPs), just as VOR and NDB determine present position from the intersection of LOPs based on radials and bearings. LORAN-C is different from VOR and NDB however, in that a LORAN-C LOP is not a straight line. A LORAN-C LOP is a curved line—specifically, a hyperbolic curve, or hyperbola for short. (Figure 9-2 shows what hyperbolic LOPs look like.)

The reason LORAN-C LOPs are hyperbolas, and not straight lines, is because of the way LORAN-C works; the fact that LORAN-C LOPs are hyperbolas and not straight lines affects the way in which LORAN-C determines position, and is the source of certain limitations. It is therefore important that LORAN-C users be somewhat familiar with hyperbolic LOPs.

LORAN-C requires a network, or chain, of at least three stations to determine a fix, one of which is designated the Master station, while the others (a minimum of two) are designated Secondaries. The Master sends out a continuous string of pulses in the Low Frequency band (100 KHz), which in turn triggers the Secondaries to send out similar signals. The LORAN-C receiver decodes all the signals, separates Master and Secondaries (based on coded differences that identify each), corrects for the amount of time

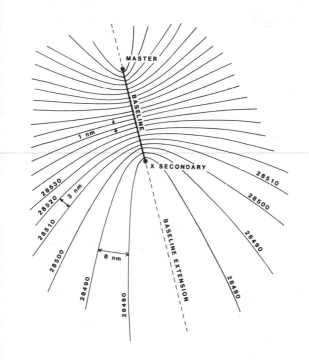

Fig. 9-2. LORAN-C uses hyperbolic Lines of Position to determine present position. Illustration courtesy of II Morrow, Inc.

a Secondary signal 28,500 microseconds later, then I must be somewhere along a line where those two signals will always be received 28,500 microseconds apart," and it turns out that that kind of line is a hyperbola.

It is easy to understand, intuitively, that the LOP formed by a radial or bearing is a straight line. It is not so easy to understand how an LOP formed by all the locations where the time difference between the receipt of two signals transmitted simultaneously is a hyperbola. It can perhaps best be seen graphically (Fig. 9-3). Each of the points depicted, P_1, P_2, P_3, and P_4, are different distances away from the Master (M) and Secondary (S) stations, but each is the same difference in distance: 200 nautical miles. For instance, P_1 is 100 NM from S and 300 NM from M, which is a difference of 200 NM. P_3 is 300 NM from S and 500 NM from M, quite a bit further away, but the difference in distance between M and S is still just 200 NM. Looking at the other curve, P_4 is 400 NM away from S and 200 NM away from

it took the Secondaries to be triggered, and then subtracts the difference between receipt of the Master signal and each of the Secondaries. If a LORAN-C equipped aircraft is the same distance from the Master and a Secondary, for instance, then the two signals will be received simultaneously; if it is closer to the Master, then that signal will be received first by an amount that will be direct proportional to its displacement from the mid position (and vice versa if it is closer to the Secondary). Since radio signals travel at the speed of light (186,000 statute miles per second), these measurements can be used to translate the differences in time directly into differences in distance. (Time multiplied by the speed of light equals distance.)

LORAN-C therefore deals with the difference in time between the receipt of two signals transmitted simultaneously (after correcting for the time it takes the Master signal to reach the Secondary). The LORAN-C receiver/computer says, "If I received a Master signal at exactly this moment, followed by

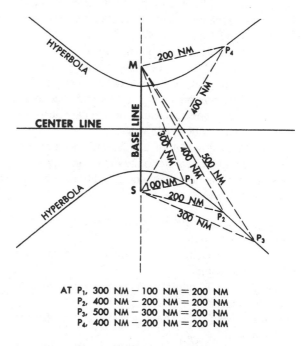

AT P_1, 300 NM − 100 NM = 200 NM
P_2, 400 NM − 200 NM = 200 NM
P_3, 500 NM − 300 NM = 200 NM
P_4, 400 NM − 200 NM = 200 NM

Fig. 9-3. The difference in distance between Point M and Point S is the same at any point along either of these two hyperbolas: 200 nautical miles.

M, also a difference of 200 NM. The same can be said for any point plotted and measured along either of the hyperbolas depicted—they are all the same (absolute) difference in distance from M and S—200 nautical miles in this case. (LORAN-C uses difference in time to get difference in distance, but the end result is exactly the same.)

Note that two possible hyperbolas result from the same difference in distance. The LORAN-C receiver can tell the difference between the M signal and the S signal, which means that it can determine which signal arrived first. As long as it knows which signal arrived first—the Master or the Secondary—then it can also tell which of the two hyperbolas is the correct one and eliminate the other. The end result is a single hyperbolic LOP based on the difference in time between the receipt of two simultaneously transmitted signals.

By repeating the calculation with the same Master but with a second Secondary signal, a second hyperbolic LOP can also be determined. The intersection of those two hyperbolic LOPs results in present position. The built-in computer then converts that fix into a lat/long coordinate (a function formerly performed by busy navigators).

It isn't necessary that you understand exactly why a LORAN-C LOP is a hyperbola. It is important that you understand that LORAN-C determines position from the intersection of two LOPs and that those LOPs are hyperbolas, because these characteristics affect the way LORAN-C works.

Most LORAN-C units use the built-in computer to provide a great deal of additional information besides present position. Some of the additional bits of information provided are:

Bearing (BRG): Direction from present position to a selected waypoint.
Range (RGE): Distance from present position to a selected waypoint.
Track (TRK): The actual course made good; the path over the ground.
Ground Speed (GS): The speed of the aircraft over the ground.
Estimated Time En Route (ETE) or Time-To-Go (TTG): The estimated time remaining, assuming no change in ground speed, to reach a selected waypoint.

Cross Track Distance (XTD): The straight line distance between the aircraft track and the desired track; the distance "off course." Direction, right or left, is also included.
Track Angle Error (TKE): The difference, in degrees and direction (right or left), between the actual aircraft track and the desired track.
Desired Track (DTK): The Great Circle course between selected waypoints.
Winds: Calculated True wind direction and velocity in knots.
Drift Angle (DA): The difference, in degrees, between aircraft heading and track.

Figure 9-4 graphically illustrates the relationship between desired track and many of the parameters described above.

Not all units provide all of this information, and the calculation of some of this information requires the entry of TAS and Heading; however, even the least expensive, VFR-only units provide more than present position. This is an enormous advance over earlier LORAN systems, which not only did not provide any additional information, but required the pilot to manually chart intersecting hyperbolas just to determine present position.

AREAS OF COVERAGE

Since LORAN-A was primarily a marine navigational system, it was assumed by those responsible for developing and implementing LORAN-C that LORAN-C would simply be a more accurate, longer range, easier to use marine navigational system. Even though most of the major avionics manufacturers now produce LORAN-C systems, LORAN-C's nautical origins can still be seen in certain ways. One of the most obvious ways is LORAN-C's continuing orientation toward over-water areas. The dotted lines in Fig. 9-5 show the ground wave limits for LORAN-C. (Extended, sky wave coverage is sometimes available outside of these limits.) Most of the existing chains are located along one coast or another, or on islands, providing fairly complete North Atlantic and Pacific coverage.

Fortunately as you can also see from Fig. 9-5, LORAN-C does provide some coverage of land mass

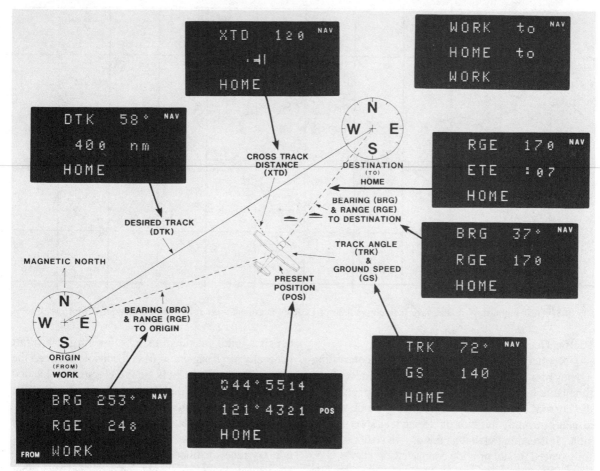

Fig. 9-4. This diagram illustrates the relationship between the desired track, aircraft position, cross track distance, and track angle. Various LORAN-C displays are shown in the dark boxes. Illustration courtesy of II Morrow, Inc.

areas as well, however a significant gap exists in the mid sections of the United States (commonly called the "mid-continent gap"), and coverage is lacking over nearly all of Mexico, the Caribbean, and a good part of Europe. No coverage currently exists at all in South America or Australia, and very little exists in Africa or Asia. The mid-continent gap has inhibited LORAN-C's development in the U.S. for private aviation, and the lack of worldwide coverage has limited LORAN-C's capability as an international long range navigational system; most aircraft regularly flown internationally rely on either OMEGA/VLF or INS, or both, for their long range navigation.

The virtues of LORAN-C are such that many general aviation pilots have installed LORAN-C sys-

tems anyway, but the closing of this gap (a very high priority among the LORAN-C manufacturers, as you might guess) is essential to general acceptance of LORAN-C in the NAS, and the extension of the chain system to include other areas of the world is essential for international acceptance. For pilots currently based in a band that runs from North Dakota to western Texas and New Mexico, and who fly mostly within that general area, LORAN-C will be of limited utility; for all others, this gap will limit the utility of their LORAN-C systems to one degree or another, depending on the type of flying they do, but for most it should not represent a serious limitation. For international operations, LORAN-C has some serious gaps, although coverage over the North Atlantic and

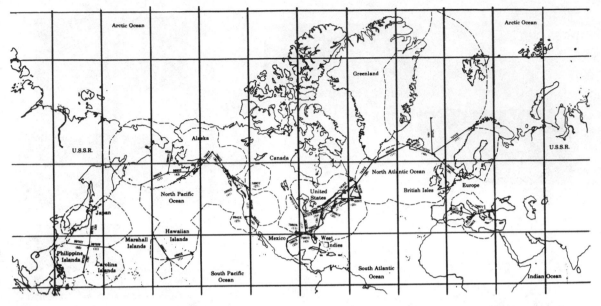

Fig. 9-5. Areas outlined by dashed lines indicate the limit of LORAN-C ground wave reception. Courtesy of II Morrow, Inc.

Pacific Oceans is satisfactory.

Neither the U.S. Coast Guard (the controlling agency) nor the FAA will commit themselves to a time table for closing the mid-continent gap, although the current FAA Administrator has gone on record as being strongly in favor of closing it as soon as possible. It is estimated that as few as five stations, carefully placed, could provide complete Continental U.S. coverage. The creation of additional chains outside the U.S. has a lower priority, but may possibly come about as LORAN-C navigation becomes more common (assuming NAVSTAR/GPS, covered in Chapter 12, doesn't overtake it).

CONTROL-DISPLAY UNITS

There may be more variation in Control-Display Units (CDUs) for LORAN-C than there is for any other single navaid type. (No doubt a good part of the reason for this variation is that many of the early aviation LORAN-C units were derived from marine models by companies new to aviation.) It is, however, possible to make certain generalizations about LORAN-C Control-Display Units. One such generalization is that the units at the lower-priced end of the market tend to be panel mounted, while more ex-

pensive units intended mainly for larger aircraft generally are designed to fit in a console between the pilot seats (with the bulk of the electronics located in a remote unit). Another generalization is that the less expensive models tend to use light emitting diode (LED) data displays in a dot-matrix format, while more expensive units tend to use either CRT (cathode ray tube) or fiber optic displays for data. More sophisticated units also tend to be patterned after other existing long-range systems, such as OMEGA/VLF and INS. In fact, it can be difficult to distinguish between a sophisticated LORAN-C CDU, and an equivalent OMEGA/VLF or INS CDU—the keyboards, displays, and mode selections tend to be virtually identical. An example of a panel mounted LORAN-C CDU is shown in Fig. 9-6. A console mounted CDU is shown in Fig. 9-7.

Since CDUs vary so much from manufacturer to manufacturer (and even from unit to unit within a single manufacturer), no attempt will be made here to describe operating procedures for any given LORAN-C unit. The manuals supplied with each unit should be consulted for information on specific operating procedures. General information regarding setup, chain selection, and navigational technique is, however, included in the next section.

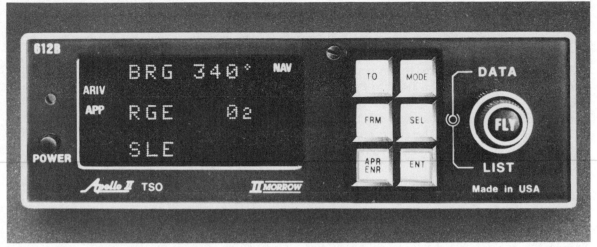

Fig. 9-6. A typical panel mounted LORAN-C unit. Photo courtesy of II Morrow, Inc.

INITIALIZATION

Most long range navigational systems have to be "initialized" prior to use. That is, unlike a VOR or an NDB system that can simply be turned on and used, most long range systems require that certain procedures be accomplished before the system can be put into use. Initialization requirements vary with the type of navigational system, but they usually involve providing the system with a starting point, selecting and capturing the appropriate facility, and, in the case of inertial systems, allowing the unit time to align its gyros. Initialization for LORAN-C is not difficult or time consuming, but it is essential to proper functioning.

There are currently a total of 15 chains (a chain is a network of Master and at least two Secondary stations) worldwide, eight of which are in North America. Most LORAN-C units (including all of the less expensive units) require that the pilot manually select the specific LORAN-C chain to be used. In some cases, the choice of chains will be obvious, since only one chain will provide coverage in the departure area. In other cases, coverage areas from one chain to another may overlap. In these cases the pilot must select the chain which appears to provide the best initial coverage. Manufacturer-supplied charts are available to help in the selection process.

Chains are identified by their *Group Repetition Interval* (GRI). The Group Repetition Interval is the amount of time, in microseconds, between the start and stop of the Master station transmission. Each station has a slightly different interval, which enables the LORAN-C receiver to electronically identify the Master station. It is also used by the pilot to tune the Master station. For instance, the GRI for the Northeast U.S. Master station is 99,600 microseconds. The

Fig. 9-7. A console mounted Control-Display Unit for a sophisticated LORAN-C system. The receiver/processor for this type of unit is housed in a separate, remote mounted unit. Photo courtesy of Offshore Navigation, Inc.

identification for this station (you might think of it as the frequency, but it isn't), is 9960; entering that number into the LORAN-C unit will tell the receiver to search for and use that chain initially for navigation. (Some manufacturers drop both zeros from the actual GRI time interval: 99,600 microseconds becomes a GRI ident of 996. It might be easier if chains were identified by channel numbers or letters, and it certainly would be easier if there were greater standardization among manufacturers, but this is not the case.)

Once a GRI has been selected, the unit then goes into a search, identify, and track cycle. During this cycle the unit attempts to receive the Master station for the GRI selected, and then looks for the two best Secondary signals in that chain. Once it has all the signals it needs, it electronically examines the shape of the individual pulses, looks for the optimum timing point in the cycle, and once that point has been established for each of the signals, it begins tracking. (Tracking, in this case, means the process of timing the differences in arrival time of the signals and then converting those time differences to hyperbolas—LOPs.)

Once tracking has begun, the initialization is complete and the unit is ready to provide position information and begin computing course and speed information. For most systems, this initialization process takes from two to five minutes after selection of the GRI. Initialization does not have to be done on the ground, or even over a known point (as some long-range systems do), but it must be completed prior to using the system for navigation; therefore, if the pilot wants to be able to use the LORAN-C system immediately after takeoff, it should be done as part of the pre-taxi checklist.

Most units also require the pilot to manually change GRIs as the aircraft travels from one area of coverage to another. As the aircraft reaches the fringe areas of coverage for one chain, the GRI for the next chain along the route of flight should be entered into the receiver. The receiver will then re-compute present position and begin again the process of computing all derivative information based on the new chain. If this changeover is not made, the position and speed reports will become increasingly inaccurate, and

eventually one of the necessary signals will be lost entirely and the display will revert to inactive status.

Some of the more sophisticated units provide for automatic selection of the optimum chain, and no pilot input of GRIs is necessary either for initialization or en route use (although manual override is usually provided). This adds significantly to the cost of the unit, but it is an important feature for operations which frequently rely on LORAN-C for navigation overwater where other sources of back-up navigational assistance are not available. For the average general aviation pilot who normally flies in the area covered by one or two chains (the entire eastern half of the United States can be covered by two chains, for instance), this level of sophistication is not necessary.

VFR LORAN-C NAVIGATION

LORAN-C can be used to great advantage by VFR pilots for primary route guidance, supplemented by VOR information, pilotage, and dead reckoning as necessary for verification and back-up. For a relatively small expense, LORAN-C provides point-to-point navigational guidance with great reliability, accuracy, and ease of use.

While specific procedures vary from unit to unit, point-to-point VFR navigation using LORAN-C is essentially simplicity itself: The point of origin is entered into the unit by one of several methods—manual entry of lat/long coordinates, the selection of a stored airport identifier or waypoint, or by selecting position mode and allowing the LORAN-C to determine present position itself. Once present position has been entered, the destination waypoint is then entered either by lat/long coordinates or by airport identifier. At this point the unit can be switched to navigation mode and course guidance will then be reflected on a built-in CDI, a remote CDI dedicated to the LORAN-C system, or on a VOR CDI or HSI.

Cross Track Distance. (XTD) is a term frequently used in long-range area navigation. It refers to the distance between the Desired Track (DTK) and the actual track (TRK). Cross track distance is the actual distance the aircraft is off course. Cross track distance can be used directly to intercept the desired course—LORAN-C course guidance does not have to be routed to a CDI. If, for instance, the LORAN-C unit indicates a Cross track distance 13 nautical miles left of course, the pilot can simply correct to the right and monitor the XTD—if it decreases, then he knows that the new heading is succeeding in bringing the aircraft back to the Desired Track. When the XTD reads zero, the aircraft is back on course and the pilot can then adjust the heading to stay on course. By monitoring the XTD and making changes in heading as necessary, the course can still be bracketed and a wind correction angle established without having to use a CDI.

This is the most direct method to track using LORAN-C, and while most pilots will probably still prefer to use the more familiar CDI, it is good to know that XTD can be used directly to track to the destination should the CDI fail. Whatever the particular method of tracking, the result is the same: direct, Great Circle navigation, corrected for wind, from any one point to any other within the area of LORAN-C coverage.

If you become confused at any time by the displays, abbreviations, or distinctions among courses, bearings, tracks, and cross tracks (like any new system, proficiency comes with experience), remember that all LORAN-C actually does is calculate present position—the rest is all derivative information. This means that at any given moment you can ask the LORAN-C to tell you where you are, and plot that position on your chart. You can then take up an estimated heading from that position to your destination, and in a few minutes ask it again for your position. Then plot that position to see how you are doing: if it looks like you are on course, fine, but if you are still not headed exactly in the right direction, then take up a new heading that looks like it will take you in the direction you want, and repeat the process.

In this way, by monitoring your position with the LORAN-C, and making corrections as necessary to proceed in the right direction, you can obtain most of the benefits of LORAN-C while avoiding errors due to confusion or misunderstanding about displays and terminology. This may result in a somewhat zig-zag track as you plot positions and estimate headings, but it is an easy way to learn how to use the system by building on what you already know.

IFR APPLICATIONS FOR LORAN-C

LORAN-C offers no advantage under Instrument Flight Rules over any other area navigational system in obtaining random route clearances—essentially the same limitations and procedures must be followed to obtain direct routings using LORAN-C as must be observed for VOR/DME-based RNAV. (Refer back to Chapter 5, VOR/DME-Based RNAV, for a listing of these requirements.) The requirement to file at least one waypoint per en route center is dropped for aircraft filing random route flight plans based on lat/long coordinates, but the random route portion is restricted to Flight Level 390 and above—a disqualifying restriction for most aircraft. (Even many turbojet-powered aircraft cannot climb directly to FL 390 at normal takeoff weights). Within the NAS, LORAN-C is essentially restricted to a supplemental role at the present time.

Supplement to Airway Navigation

In its supplemental role, LORAN-C provides many of the same advantages and has many of the same capabilities that VOR/DME based RNAV does for normal airway navigation, however LORAN-C tends to be more accurate and is generally easier to use. LORAN-C can be used to: improve airway tracking, obtain direct clearances to distant fixes (VORs, outer markers, airway intersections), assist with visual approaches, provide distance to touchdown, aid in NDB tracking, provide a degree of airway cross-checking capability. (Again, refer to Chapter 5 for a complete discussion of these supplemental applications.)

In addition to the supplemental IFR applications common to both LORAN-C and VOR/DME-based RNAV systems, LORAN-C can be used in several additional ways to supplement normal airway navigation. (Not all LORAN-C systems have all of these capabilities, but most have some of these capabilities.)

Course Off-Sets

A course off-set is a course that is parallel to the desired course; i.e., both courses have the same heading, but are separated by a given distance. (A very few VOR/DME-based RNAV units have course off-set capability, but this is an exception to the rule; in general, course off-set capability is available only with long-range systems.)

The advantage to a course off-set is that it enables the pilot to fly around certain areas such as areas of hazardous weather, Restricted Areas, high terrain, even slower traffic, without losing navigational guidance—the original DTK is retained, as is all range and bearing information to the destination. As soon as the area to be circumvented is passed, the unit can be switched back to normal navigation, and either the original DTK re-intercepted, or a new DTK established from that point.

Ground Speed, Distance, and Time Remaining

Probably one of the greatest advantages to LORAN-C as a long-range navigational system is its ability to provide ground speed, distance remaining, and time remaining at all times, completely independent of DME facilities. While a few VOR/DME-based RNAV units do provide ground speed information (as long as the aircraft is navigating directly to or from a waypoint), LORAN-C provides ground speed at all times and without regard for whether the aircraft is headed directly to or from a waypoint or DME. This is because LORAN-C computes ground speed from changes in position, rather than relative to a ground-based facility or waypoint. The typical LORAN-C unit re-computes both present position and ground speed twice per second. With continuous position information and ground speed available, it is a simple matter for it to also compute distance remaining and time remaining.

Having this kind of information available at all times is a great help to the pilot operating on an IFR flight plan, even if he is still restricted to airway navigation. It means he will have accurate ground speed and time remaining read-outs at all times, even when being radar vectored (rather than whenever he happens to be navigating directly to or from a VOR/DME), and it will be free from slant-range error. In addition, distance remaining and time remaining can be obtained in relation to the destination, rather than just to the next VOR/DME (or VOR/DME-based RNAV waypoint). In simplest

terms, LORAN-C fills in the navigational gaps left by VOR/DME.

Long Range and Over-water Applications

While the many benefits of LORAN-C are available to all pilots, long-range systems such as LORAN-C are essential for pilots who routinely operate long distances, in isolated areas lacking other navaids, or over water. Navigation based solely on VOR and NDB facilities, backed-up by dead reckoning, can be accomplished in isolated areas, but long-range systems are so much more accurate and so much more reliable that for routine operation over water and in isolated areas, LORAN-C (or something like it) is a virtual necessity.

Over-water navigation is covered in some detail in Chapter 13. Our concern here is not so much with how to use LORAN-C over water as it is in knowing that the problems inherent in over-water navigation are best solved by a long-range system such as LORAN-C. LORAN-C changes from a supporting to a primary role when the area of operation changes from land to water. For the pilot who even occasionally contemplates operating out of range of VOR and NDB facilities, the availability of accurate position and ground speed information provided by a long-range system such as LORAN-C is extremely important.

The pilot who routinely operates IFR over long distances, but not necessarily over water or in isolated areas (coast to coast, for instance), may also find LORAN-C to be a worthwhile navigational system to have available. Not only will he often find that his navigational workload is reduced with the LORAN-C system, but he may also find that the time and fuel saved by being able to operate even part of the way along direct Great Circle routes (off airways) is substantial, significantly reducing and possibly even eliminating the cost of the system itself.

LIMITATIONS

Compared to some other navigational systems, such as NDB or DME, LORAN-C is relatively free of limitations, but those that it does have are significant and should be well understood. An understanding of its limitations is particularly important when

LORAN-C is being used over water as a primary source of navigational information. (Ironically however, accuracy and reliability are more likely to be limited when the system is used over land.)

Areas of Coverage

The single most limiting factor in LORAN-C navigation is its limited area of coverage. The manufacturers of LORAN-C equipment supply detailed charts showing prime and fringe areas of LORAN-C coverage, usually in booklet form, and these should be consulted by those interested in a precise description of areas of coverage. Pilots should be aware of the limits of LORAN-C coverage, and either plan their flights to stay within the confines of expected coverage, or ensure that at least one other source of navigational guidance is available.

Sky Waves

LORAN-C is a Low Frequency system, which means that the sky wave—a pulse that has been refracted by the ionosphere back to earth—will sometimes be received by the LORAN-C unit in addition to the normal (and preferred) ground wave. Since the sky wave travels a greater distance than the ground wave, it takes a longer period of time to reach the receiver, which adversely effects the accuracy of the plotted LOP. This is normally not a problem at distances less than 1,000 nautical miles from the furthest Master or Secondary station—the ground wave usually dominates at these distances, and most LORAN-C receivers have circuits to enable them to differentiate between the stronger ground wave and the weaker sky wave. At greater distances, however, usually over 1,400 nautical miles (although the exact distance will vary with terrain and time of day), the ground wave becomes so weak that the sky wave begins to dominate. Most LORAN-C receivers can also detect this condition and have circuits to compensate for the additional distance traveled by the sky wave—accuracy is reduced slightly, but it is generally acceptable, and in any case is usually more accurate than dead reckoning.

The problem comes at the intermediate distances—1,000 to 1,400 nautical miles. At these distances the sky wave and ground wave are about equal, which confuses the receiver and distorts the results. Some LORAN-C units avoid the problem of sky wave distortion by automatically rejecting all sky waves. This solves the problem of ambiguity at the intermediate distances, but reduces the operating range of the system—a cost effective trade-off that maximizes reliability at the expense of range. Others attempt, as best they can, to sort out the two. Systems that use both the ground and the sky wave can certainly be safely used at these intermediate distances, but this limitation should be clearly understood and a certain amount of healthy skepticism should be applied to any position fix that falls within this intermediate range, particularly at night.

Precipitation Static

As an aircraft passes through precipitation, an electrical charge builds on the metal skin of the aircraft. This charge affects all electronic systems to a certain extent, but it affects long range, low frequency systems the most. This charge can seldom be entirely eliminated, but it can be reduced by airframe grounding and by installing wicks—metal probes which extend into the airstream to help in dissipating the static charge.

Precipitation static causes electronic "noise." Most LORAN-C receivers constantly monitor the signal-to-noise ratio (SNR), and issue a warning when the ratio drops below a certain minimum level. Since SNR is a ratio—signal divided by noise—a low SNR can be caused by either a low signal level (out of range), or a high noise level (precip static), or a combination of each. The practical effect of precipitation static is therefore the same as a decrease in the signal level, and that is a reduction in range. The only solution available in flight is to leave the area of precipitation. If this cannot be done, then the LORAN-C information will not be available until the SNR improves. (If this regularly occurs in prime coverage areas, the aircraft may require additional grounding and wicking.)

Additional Secondary Factor

LORAN-C was designed as an over-water system, and all of the assumptions as to signal speed and transmission are based on signals propagated over sea water; calculations based on signals propagated over land will be slightly in error. Most modern LORAN-C receivers have compensating circuits, called *Additional Secondary Factor* (ASF) circuits, to correct for this condition. Some even allow the ASF circuits to be modified by the pilot when a discrepancy is noted in the position read-out over a known geographic lat/long fix. These propagation errors are usually not significant and only affect the accuracy slightly and within predictable limits. The pilot should be aware however, that peak accuracy can only be obtained with LORAN-C when operating over sea water, using coastal chains.

ACCURACY

Just as VOR navigational guidance varies in accuracy with distance from the station and with the geometry of the radials, so does LORAN-C accuracy vary with distance and LOP geometry; however, LORAN-C is more accurate than VOR to start with, and it retains that advantage at even its maximum range and least favorable crossing angles. Generally speaking, LORAN-C will be more accurate at 1,000 nautical miles than VOR will be at 20 nautical miles.

Repeatable Accuracy

LORAN-C accuracy is typically measured in two different ways (and when you read system specifications, it is important that you compare similar types of measurements—the differences are substantial). The first measurement is called *Repeatable Accuracy*. It refers to the system's ability to return to a specific previous position. This is much easier for a LORAN-C system to do than it is for it to determine its position in space. (The main reason for this is that the terrain and propagation errors unique to that particular location will be essentially the same upon return, and this effectively eliminates these factors as sources for error.) Typical repeatable accuracy for LORAN-C is 0.01 nautical miles; that is, after logging a known position, say the takeoff end of the runway, the system should be able to return to that same point to within 0.01 nautical miles, or about 60 feet.

Absolute Accuracy

The other LORAN-C measurement of accuracy is *Absolute Accuracy*, or the ability to determine present position independently. This accuracy will vary from 0.1 nautical miles to as much as 2.5 nautical miles, depending on distance from the station, geometry of the crossing angles, design choices made by the manufacturer, terrain and environmental conditions, and the signal-to-noise ratio. At the upper levels of sophistication (and cost), worst-case absolute accuracies of 0.5 miles or less are possible. Each of these accuracy factors is discussed in the next few paragraphs.

Distance. Distance from the LORAN-C station affects accuracy by increasing the distance between LOPs. (See Fig. 9-2). LOPs crossing the baseline (the imaginary line between Master and Secondary) at the halfway point tend to be close together at all distances, while LOPs crossing the baseline close to either station fan out as distance increases from the station, making those LOPs generally less accurate than LOPs crossing the midpoint. The area behind either the Master or the Secondary (near what is called the baseline extension) is the area of greatest inaccuracy, and LORAN-C position information is unreliable and should not be used along the baseline extension itself. (A good way to remember this is to think of the baseline extension area as a *shadow area*—an area without coverage due to the shadow effect of the station. Accurate position information can sometimes be obtained in these areas by switching to another chain).

Geometry of Crossing Angles. The angle at which the LOPs intersect each other also affects accuracy. LOPs that cross at 90 degree angles produce small, square areas of position between hyperbolas, while LOPs that cross at oblique angles produce large, diamond shaped areas between hyperbolas. (See Fig. 9-8.) The geometry of the crossing angles is largely outside of the control of the pilot and depends upon his position in relation to the Master and Secondary station. It is, however, one of the more important factors affecting the absolute accuracy of the fix.

Design Choices. The LORAN-C receiver calculates the time difference between Master and

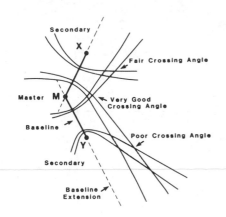

Fig. 9-8. LORAN-C accuracy varies from place to place due to variations in LOP geometry, or crossing angle. Illustration courtesy of II Morrow, Inc.

Secondary signals. The computer converts that time information into distance information, and then plots a position in terms of latitude and longitude. This is an extremely complicated process involving a large number of variables, and the formulas used (called algorithms) vary from manufacturer to manufacturer. Different receivers will therefore sometimes produce slightly different lat/long coordinates from the same set of hyperbolic LOPs, and each type of receiver will be more accurate in some situations than in others (depending on the choices made by the engineers who developed the algorithm). These kinds of design choices account for a small part of the variability in LORAN-C accuracy from one location to another.

Terrain and Environmental Factors. Terrain and environment affect the strength of the ground and sky waves, which in turn affects the accuracy of the fix. The most accurate fixes are based on ground waves alone. Fixes based exclusively on sky waves will be a little less accurate, and fixes based on a combination of ground and sky waves are the least accurate. Sky waves are strongest at night. Terrain reduces the strength, and therefore the range of the ground wave, and signals traveling over land will be less accurate, in general, than signals that travel exclusively over water, despite compensating circuits (ASF). All of these factors combine to make LORAN-C navigation most accurate over water, during the day, and least accurate over land, at night.

Signal-to-Noise Ratio. As the SNR drops, ac-

curacy decreases. Signal-to-noise ratio is an indirect function of distance from the station: the farther away, the lower the SNR; therefore, the farther away, the less accurate the fix. Most units issue a warning when the SNR reaches an unacceptable level.

All of these factors affect the absolute accuracy of LORAN-C, and affect it in variable and somewhat unpredictable amounts. Nonetheless, the accuracy even at the worst case limits is still phenomenal: absolute accuracy no worse than three nautical miles at distances as great as 2,800 nautical miles, and repeatable accuracies of 0.01 nautical miles. In most cases LORAN-C is, in fact, more accurate than other, more expensive long-range systems such as OMEGA/VLF and INS (however, they retain certain other advantages which negate their disadvantage in accuracy and cost).

GROWTH PROSPECTS

In just a few years, LORAN-C has grown from obscurity to perhaps the most significant new system of air navigation for general aviation since the VOR system. The only serious limitation to complete acceptance of LORAN-C in the U.S. NAS is the bridging of the mid-continent gap, and that gap should be closed soon.

LORAN-C has been demonstrated on an experimental basis as a non-precision approach aid, and it is probably only a matter of time before fully certified LORAN-C approaches begin to appear in the approach manuals. (An example of an experimental LORAN-C approach is illustrated in Fig. 9-9.) When LORAN-C approaches do appear, they will almost certainly be more accurate than VOR or NDB approaches (but they will still be non-precision approaches, and as such will probably continue to be limited to minimums no lower than 400 feet AGL and 3/4 miles visibility). LORAN-C approaches will not require the purchase or installation of additional ground facilities—it will only be necessary to design and flight test the approach procedure itself, since the LORAN-C facilities already exist. If LORAN-C were used for nothing more than to retire the existing NDB approaches and lower the minimums for many VOR approaches, that alone would justify the system for most pilots.

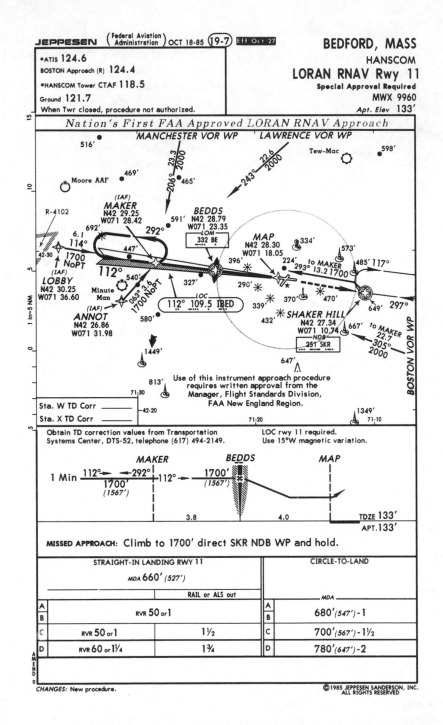

Fig. 9-9. The first FAA approved LORAN-C approach was established at Bedford, Massachusetts, HANSCOM airport (near Boston). LORAN-C approaches are experimental at the present time and require special authorization, however, this is the first step towards full development of LORAN-C as an approach aid. Reproduced courtesy of the FAA and Jeppesen Sanderson, Inc. Not actual size. Not to be used for navigation. NOT FOR COCKPIT USE.

Long-range navigational systems have traditionally been priced well outside of the acceptable range for the average general aviation pilot: OMEGA/VLF systems are priced from $19,500 to $68,000 (1986); INS systems start at $100,000. LORAN-C on the other hand, is an affordable system for the average general aviation pilot: fully IFR-capable LORAN-C receivers are available at prices that compare very favorably to panel-mounted VOR receivers—$3,000 to $4,000. With increased acceptance and production, prices may decrease even further. While LORAN-C is still a supplemental system (from the FAA's point of view), the fact that it can be purchased for about the cost of an additional VOR is a tremendous breakthrough for general aviation.

What is perhaps most promising about LORAN-C is that it appears that LORAN-C and NAV-STAR/GPS (a proposed satellite navigation system designed to provide three-dimensional accuracies to within 150 feet at very reasonable cost to the user) are perfectly complementary. Specifically, NAV-STAR/GPS signals can be used as a time standard to calibrate LORAN-C pulses, resulting in even greater LORAN-C accuracies than are possible today, while LORAN-C is the perfect back-up for those periods each day when the satellite system, because of less than perfect positioning, is least accurate. (NAVSTAR/GPS is covered in Chapter 12.) It is still too early to know for sure what will develop with these two systems, but it is very possible that 10 years from now the typical fully equipped IFR aircraft will not be described as "Dual VOR, DME, NDB-equipped," but will instead be described as "DUAL LORAN-C/NAVSTAR-equipped."

For all of these reasons it appears that LORAN-C's future is very bright, especially if the system can be expanded to include all of the U.S. and more of the rest of the world—which isn't bad for a system that was never intended to be used for air navigation in the first place.

Chapter 10

OMEGA/VLF Navigation

OMEGA/VLF is a long-range, random-route navigational system based on Very Low Frequency (VLF) transmissions from a small number of very powerful transmitters, each capable of providing signal coverage at distances as great as 11,000 nautical miles. OMEGA/VLF is an enormously complex system in nearly every sense—electronically, physically, politically—and it is somewhat vulnerable to atmospheric distortion, but it is also a very capable navigational system having certain unique characteristics. It fulfills an important role in the overall navigational scheme, providing primary and secondary route guidance for many aircraft in daily operations, and serving both as a primary and as a supplemental en route system in international operations.

OMEGA/VLF is a modern system of navigation, one that is based on advanced technologies (computers, atomic clocks), yet it is also a mature system, having been proven in daily operation for over a decade. It is not an inexpensive system from the user's point of view, although the cost is generally less than half that of inertial systems to which OMEGA/VLF is frequently compared.

OMEGA/VLF is neither the most nor the least accurate navigational system (not in all situations at any rate), and while it is not difficult to use, it is sophisticated and powerful enough to require a relatively high degree of training and proficiency to be used properly. OMEGA/VLF is, in other words, a system that balances many different factors in order to arrive at something that is unlike any other system.

Perhaps the most interesting aspect of OMEGA/VLF navigation is that it combines principles of dead reckoning with ground-based electronic positioning to form a composite system that uses both electronics and dead reckoning to achieve accuracies and speeds not possible with either separately. OMEGA/VLF has been described as a dead reckoning system that continuously updates its calculations with position information, although it is more commonly described as a positioning system that improves the quality of its position estimates with dead reckoning—there is some truth to both points of view. In short, while OMEGA/VLF appears, on the surface, to be very similar in many ways to other systems, on closer examination nearly everything about

OMEGA/VLF is unique. It is a very interesting system.

Because OMEGA/VLF is so complex, this chapter is organized a little differently from preceding chapters. While the basic principles of OMEGA/VLF (actually two different systems) will be covered, no attempt will be made in this chapter to describe in detail exactly how the system does what it does—OMEGA transmits on four different frequencies, and there are five different methods of deriving valid position information from OMEGA transmissions (using different numbers of stations and different parts of the signals transmitted). VLF signal interpretation is done in yet another way, and the merging of the two systems into what appears to be a single unit is a subject in and of itself. OMEGA/VLF is a very technical system, and will be described only in general terms.

Those principles of operation that do affect the operation and limitations of the system will of course be discussed; pilots with engineering or science backgrounds who would like to know more about the theory of OMEGA/VLF operation will find much to keep them occupied in various issues of *Navigation,* the journal of the Institute of Navigation (see Appendix A). In addition, several of the OMEGA/VLF manufacturers have published useful background documents—a booklet entitled *OMEGA Navigation* by the Canadian Marconi Company is recommended for starters. These can generally be obtained, free of charge, by writing the manufacturer.

I will not go into detail, either, concerning the VFR and IFR applications for OMEGA/VLF (Great Circle routings, course off-sets, direct courses, and so on). These applications, both primary and supplemental, have already been covered in some detail in the chapters on VOR/DME-based RNAV and LORAN-C. OMEGA/VLF can do anything VOR/DME-based RNAV or LORAN-C can do in terms of route and supplemental airway guidance. In some cases its accuracy will be less than either of those systems, and in some cases it will be greater (OMEGA/VLF accuracy is discussed later), but the practical VFR and IFR applications for OMEGA/VLF are very similar to those for VOR/DME-based RNAV, and are nearly identical with those for LORAN-C, the only significant difference being that OMEGA/VLF guidance is available worldwide.

OMEGA AND VLF

OMEGA/VLF is, in reality, not one, but two systems: OMEGA and VLF. Each of these systems exists for entirely different reasons—OMEGA for navigation, VLF for communication—but both can be used for long-range navigation. It is important that pilots using OMEGA/VLF for navigation be very familiar with the fundamental elements of these two systems, in order to understand the overall OMEGA/VLF system characteristics and limitations.

VLF System

VLF (for Very Low Frequency) is a very high powered communication system operated by and for the U.S. Navy primarily for direct communication with submarines anywhere in the world. There are a total of seven VLF transmitting stations, each strategically located at various points around the globe, and each operating on a unique frequency in the VLF band. The VLF frequency band begins at 10 kHz and ends at 30 kHz, which means it overlaps with the audio range (.020-20 kHz). A VLF transmitter, therefore, physically transmits what amounts to sound through the ground, and for this reason VLF transmitters must be extremely powerful: 500,000 to 1,000,000 watts. (By way of comparison, High Frequency communications transmitters, such as those typically used by turbine powered aircraft for long-range, over-water communications, have power ratings in the 100- to 450-watt range.)

The main advantage to VLF, and the reason the Navy chose it for worldwide communications, is that given enough power and a suitably sensitive antenna/receiver, VLF frequencies can be reliably received at very long distances: 6,000 to 11,000 nautical miles. (This compares to 1,000 to 2,800 nautical miles for Low Frequency systems.) These VLF communication transmissions also happen to be excellent signal sources for navigation, since they are very stable and predictable, and provide worldwide coverage.

The problem with the VLF system is that it was never intended to be used for navigation and is not under the control of any agency having jurisdiction over navigation (such as the International Civil Aviation Organization—ICAO). The U.S. Navy does not object to its being used for navigation, but it does not support it either. The Navy does provide information on the system's status, and attempts to inform the aviation community when any element is scheduled to be shutdown for maintenance, but it does not do so through the normal NOTAM channels—it is necessary to make a toll call to receive a taped message. (The current telephone number can be found in the AIM, para 21. b. [2].) The VLF system is an independent system, operated for and controlled by the U.S. Navy, and no guarantees are intended or made as to its reliability or suitability for navigation.

Despite these shortcomings, the VLF network is too powerful and stable not to be used. Most OMEGA/VLF receivers use VLF signals on an "as available" basis—if a good VLF signal exists, then the OMEGA/VLF receiver uses it. Units vary, however, in the order in which they use the VLF transmissions: some look for VLF first, switching to OMEGA when VLF is not available, while others start with OMEGA and only revert to VLF when OMEGA accuracy drops below a certain level.

Since VLF is not a primary navigational system and can only be used to supplement other forms of navigation (when it happens to be available), its practical significance for aviation is somewhat limited. The FAA will, in fact, approve OMEGA/VLF systems for IFR en route use only if the OMEGA portion is capable of meeting all system specifications completely independently from the VLF portion. To the extent both OMEGA and VLF are often combined for increased accuracy when VLF is available, pilots generally think of these combined systems as one—"OMEGA/VLF"—which is not wrong, but it is important to understand that they are two entirely separate systems.

OMEGA System

The OMEGA system was designed for navigation, and is under the control of an agency having navigational jurisdiction, specifically the U.S. Coast Guard. NOTAM service is available on the status of OMEGA stations through normal channels (i.e., FSS), and OMEGA can be relied on for navigational guidance.

The OMEGA network works in a similar manner to the VLF network, but because OMEGA was designed exclusively for navigation, it has certain fundamental differences. OMEGA is a network of eight stations (as opposed to seven for VLF), located so as to provide worldwide coverage (see Fig. 10-1), but the OMEGA stations, because they are intended solely for navigation and not communication, transmit at lower power levels (10 kW versus 500 to 1000 kW) and have a much more complex pattern of frequencies. With OMEGA, the amount of radiated energy actually reaching the aircraft antenna can be extremely low. Consequently, antenna installation, and aircraft shielding and grounding, are very important to proper OMEGA operation.

Because the OMEGA frequency pattern is more complex than the VLF signal (a continuous, single frequency), designers have more choices in how the OMEGA signal can be utilized, which allows for certain cost/benefit trade-offs to be made: receivers utilizing the full pattern of frequencies can obtain greater accuracies, but at increased expense. Most OMEGA systems designed for air navigation use the entire signal pattern, which is one of the reasons airborne OMEGA/VLF systems tend to be relatively expensive.

OMEGA/VLF Ground Facilities

OMEGA/VLF ground facilities are themselves large, complex enterprises. Figure 10-2 shows the two types of transmitting antennas utilized for Very Low Frequency transmissions: the Valley Span type requires a valley 400 meters deep and 3,500 meters wide (over two miles wide), and the Umbrella type requires a tower 450 meters high (about 1,500 feet). Even though OMEGA transmits at lower power levels than VLF, both systems require large transmitters, each of which must be equipped with an expensive atomic clock in order that all transmissions can be exactly synchronized among all stations.

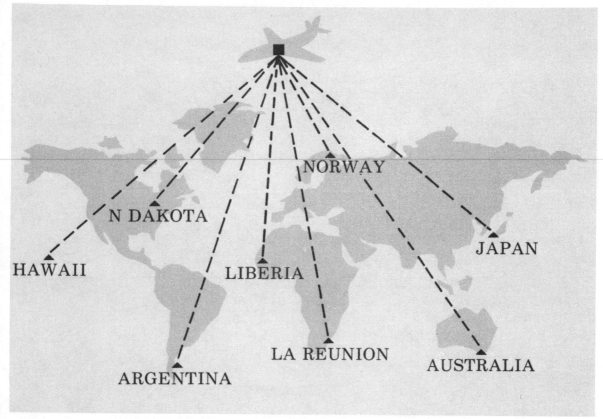

Fig. 10-1. There are eight OMEGA transmitters, strategically located around the world to provide optimum and complete signal coverage. Illustration courtesy of Canadian Marconi Company.

GENERAL THEORY OF OPERATION

Each of the eight OMEGA stations transmits a series of three frequencies: 10.2 kHz, 13.6 kHz, and 11.33 kHz. These three frequencies are always broadcast in the same order, with the same interval (0.2 seconds) between each frequency, but with a different interval after the last frequency and the beginning of the new cycle. The complete cycle for each station, including the unique interval between stop and start, lasts exactly 10 seconds and is repeated continuously. The OMEGA receiver can identify each station by its transmission pattern, however each station also transmits a fourth frequency for more positive and rapid identification.

This unique pattern is also used by the OMEGA receiver to synchronize itself with the transmitting station in order to measure the difference in phase between transmission and receipt (phase relates to the part of the cycle: beginning, middle, end). Phase differences are used to develop hyperbolic Lines of Position (similar to LORAN-C); direct phase measurements can be used to determine distance directly (similar to DME).

OPERATING MODES

There are two primary modes of determining position from OMEGA signals: the Absolute Mode and the Relative Mode. There are advantages and disadvantages to each system, and the most sophisticated OMEGA receivers can alternate between modes as the situation dictates. It is very important that the pilot understand which mode is in use, because the initialization process and the en route consequences vary from one mode to the other.

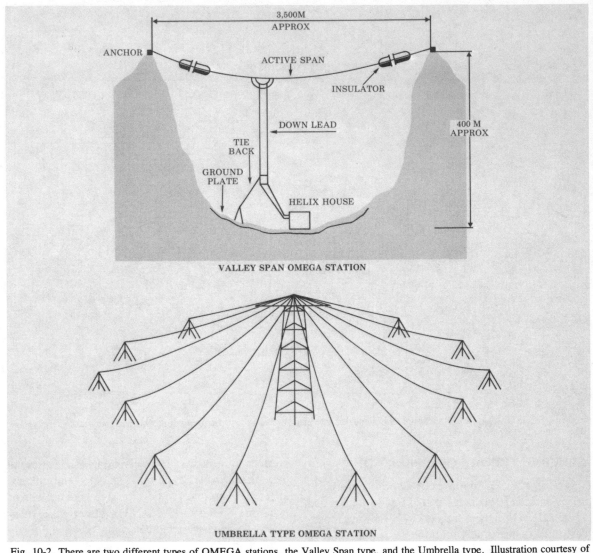

VALLEY SPAN OMEGA STATION

UMBRELLA TYPE OMEGA STATION

Fig. 10-2. There are two different types of OMEGA stations, the Valley Span type, and the Umbrella type. Illustration courtesy of Canadian Marconi Company.

Absolute Mode

The OMEGA receiver/computer can function very much like a LORAN-C system. That is, it can measure the phase differences between signals received from two stations to develop hyperbolic Lines of Position, any two of which define a fix or position. (See Fig. 10-3.) This is called the Absolute Mode, because it results in an independent determination of present position (an ''absolute'' determination), rather than one that is relative to, or dependent upon, another position. (Some manufacturers call the Absolute Mode the Hyperbolic Mode. The terms are interchangeable.)

There are several advantages to the Absolute Mode. It is fast, which means that it is not vulnerable to inaccuracies caused by high TASs. Because it works with phase differences, rather than with direct phase measurements, it is relatively free of phase induced errors, which means its ''clock'' can be relatively inaccurate without substantially affecting the results. Most importantly, its accuracy is constant with time—accuracy does not deteriorate as the flight progresses.

There is a complication, though. OMEGA actually produces multiple position fixes (in the Absolute Mode), only one of which is correct. Each fix produced, however, is in the same relative position between hyperbolic LOPs (commonly called "lanes"); present position can therefore still be determined so long as the correct lane can be identified. The OMEGA receiver (in the Absolute Mode) can determine, in other words, where it is in the lane between any two hyperbolas, but it has no way of knowing which lane it is in.

To resolve this problem and identify the correct fix, the initial aircraft position must be entered into the computer so the OMEGA receiver can identify which lane it is in to start with. After that, the computer can generally keep track of lane changes by itself. This process of keeping track of lanes is called *Lane Ambiguity Resolution.*

Lanes (the distance between hyperbolas) can be either eight nautical miles, 10 nautical miles, or 72 nautical miles wide, depending on certain design choices made by the manufacturer; therefore, the accuracy of the initial entry of position has to be accurate to within four nautical miles, five nautical miles, or 36 nautical miles, depending on the lane width chosen, to insure that the position initially entered is in the correct lane.

The choice of lane width is somewhat arbitrary. If the manufacturer decides to use narrow lanes, fewer stations at lower signal levels can be used to resolve the lane ambiguity. If the designer decides

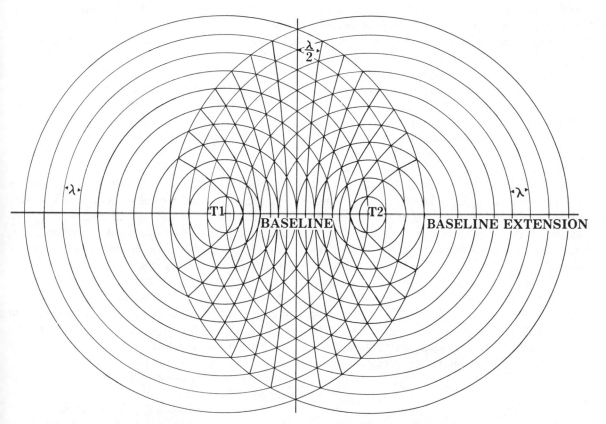

Fig. 10-3. OMEGA transmits signals in all directions, forming a pattern of concentric circles, one wavelength apart. As the wave patterns intersect, hyperbolic Lines of Position are formed. Lanes are formed in the area between hyperbolas. Illustration courtesy of Canadian Marconi Company.

to use wide lanes, less precision is required in entering initial position (plus or minus 36 nautical miles), but all three frequencies must be available from each station in use, and they must be relatively strong signals.

If wide lanes are used, and fewer than three strong frequencies are available, the receiver may skip lanes—make a mistake and end up in the wrong lane. Errors in multiples of 72 nautical miles will result. If the designer does decide to use wide lanes, circuits must be included in the design to detect those conditions where lane skip might occur and provide an appropriate warning indication.

Entry of initial position to within half a lane width is not difficult to accomplish on the ground, even for narrow lanes, since aircraft position to within a tenth of a minute of longitude and latitude—about 600 feet—can be determined from the published airport coordinates (which are located on the back of the first Jeppesen approach plate for the departure airport). If, however, it is necessary to initialize in flight, determining position to within a half a lane width may not be possible. It might be possible to determine position to within 36 nautical miles based on dead reckoning for instance, but not to within four nautical miles. If only narrow lanes can be used, then for all practical purposes in-flight initialization must be done over known fixes. If wide lanes can be used, then it may be possible to initialize with estimated fixes.

It is obvious that there are advantages and disadvantages to each choice of lane width. The important point from the pilot's point of view is that he understand that lane slip—misresolution of lane ambiguity—is possible with the Absolute Mode of position determination, and that the OMEGA receiver/computer must be initialized to an accuracy equal to or better than half the lane width in use, or serious errors in position will result. Information on lane widths and Lane Ambiguity Resolution will be provided in the operating manual, and the pilot should be familiar with the parameters necessary for proper lane resolution.

Relative Mode

The Relative Mode uses direct phase measurements to determine position in a manner similar to using distance measurements from two or more DME facilities to determine position. Using either two stations and an extremely accurate rubidium frequency standard (an atomic ''clock'') to synchronize the received signals exactly with the transmitted signals, or by using direct phase measurements from three stations, the OMEGA receiver/computer operating in the Relative Mode can measure changes in position from one fix to another—it is unable to determine exactly where it is by itself, but it can keep track of where it is if it has been told at some point where it was.

Since the Relative Mode operates on the basis of change from one position to another, the initial position must be entered into the computer with some care—any error in the initial position will be carried for the duration of the flight or until corrected. (Corrections can be made by re-initializing the system over a known point.) In addition, all errors are cumulative in the Relative Mode—the receiver does not determine a new position each cycle (as it does with the Absolute Mode), but adds or subtracts changes in position to the previous position. Errors therefore accumulate through out the flight, and accuracy, while high initially, will deteriorate as a function of time. For short flights (two to three hours), average accuracies are comparable for each mode; frequent updating can be used to maintain optimum system accuracy.

Neither mode is perfect, but both work, and each produces very similar results in practice. The most important factor for the pilot to understand is that OMEGA is capable of determining position by either of two methods, that the initialization requirements differ substantially for each, and that the results will differ somewhat depending on the mode in use. The Relative Mode is theoretically more accurate, but is dependent on precise initial entry of position and regular updates to remove accumulated errors; the Absolute Mode is less accurate initially, but is less critical in terms of initial position entry, has a relatively constant accuracy over time, and does not require regular updates.

The most advanced OMEGA receivers allow for selection of mode (Absolute or Relative), and, in the Absolute Mode, start with narrow lanes, but monitor themselves for possible lane skip. Whenever the estimated internal error appears to be in excess of four nautical miles, they automatically jump to wide lane widths. The shift from narrow to wide lanes is normally indicated to the pilot on a separate, panel-mounted OMEGA/VLF annunciator. If fewer than three strong signals are available (a prerequisite for wide lane resolution), the unit will revert to a dead reckoning mode, based on the last known position, TAS, and heading, and this will also be indicated on the annunciator.

Less complex units, for reasons of simplicity and reliability, restrict themselves to a single mode. Again, the most important part is that the pilot understand exactly how his particular unit functions, so that any unreasonable seeming fixes can be interpreted intelligently and resolved in the most appropriate fashion, and any warnings understood and proper corrective action taken.

CONTROL-DISPLAY UNITS

OMEGA/VLF is a more mature system than LORAN-C—with fewer manufacturers—and it has a narrower prime user market. As a result, OMEGA/VLF CDUs tend to be more standardized than LORAN-C CDUs, and are often patterned after INS CDUs. (INS was the first long range, random route navigational system to be widely accepted in international aviation, and as a result, newer long

Fig. 10-4. A typical OMEGA/VLF cockpit display unit. Coordinates appear along the top of the display, with messages underneath. Photo courtesy of Canadian Marconi Company.

range systems tend to pattern themselves on the INS model.)

A typical OMEGA/VLF CDU is shown in Fig. 10-4. Lat/long coordinates are displayed along the top line, and messages are displayed below that. An alpha/numeric, dual-function keypad is located on the right side, and push button data entry and data display keys are located on the left side. The receiver/computer unit (Fig. 10-5) is housed in a separate, 1/2 ATR Short box (an ARINC standard for electronic components) for remote mounting in the radio rack. The standardization of remote units makes for easier initial installation, and for interchangeability among fleet aircraft.

Figure 10-6 shows a lower cost version of the OMEGA/VLF CDU shown in Fig. 10-4. (Lower cost is, however, a relative concept: this system is currently priced at $22,440.00.) In this newer and less expensive version, the daylight readable fiber optic display of the previous unit has been replaced by a liquid crystal display, and data entry and display keys have been reduced to 12 keys in two rows. The resulting unit is smaller, allowing for easier installation in aircraft without sufficient room between the seats for

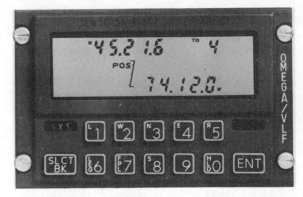

Fig. 10-6. This CDU is designed to be panel-mounted. To save space, all data entry and control functions have been reduced to two rows of keys. Illustration courtesy of Canadian Marconi Company.

a console. Since fewer keys have to accommodate more functions, the operation is slightly more complicated with this type than with the former, but the end results are the same.

Figure 10-7 shows a complete flight planning and OMEGA/VLF navigational system. With this system, flight planning and weather information can be obtained over any telephone and recorded on a 3-inch microdisk. This information is then entered into the OMEGA/VLF computer by inserting the disk in the rectangular disk drive shown (located in some convenient place in the cockpit). Flight planning and weather data can be updated en route by a radio data link between the aircraft and the central computer. The receiver/computer unit for this system is also housed in a 1/2 ATR Short box, and the CDU incorporates a CRT display that allows for up to eight lines of data.

Figure 10-8 shows yet another type of CDU. The distinguishing feature of this CDU is its color CRT display (not discernible in this black-and-white illustration), which allows for categorization of different types of data. This particular unit also includes a built-in CDI: the aircraft symbol is shown slightly to the left of course, indicating that a correction back to the right is necessary to re-intercept the Desired Track.

INITIALIZATION

Because OMEGA/VLF is a Very Low Frequency system, and because VLF transmissions travel

Fig. 10-5. As is typical, the receiver/processor units for OMEGA/VLF systems are housed in separate ARINC sized boxes for remote installation. Photo courtesy of Canadian Marconi Company.

Fig. 10-7. This is a complete OMEGA/VLF flight planning and navigational system, based on data links between aircraft and the central computer. Photo courtesy of Global Systems.

both along the ground and by repeated refractions off the ionosphere (refer back to Chapter 6, NDB Navigation, for a review of radio energy characteristics at the low end of the frequency spectrum), the height of the ionosphere, which varies both from day to night and from season to season, must be taken into account. Fortunately, these daily and seasonal variations in the height of the ionosphere (called *Diurnal Effect*) are predictable and can be permanently stored in the OMEGA/VLF computer. However to activate these corrections most systems require that the date and time (UTC) be manually entered prior to use. (Some of the newer units have built-in, battery-powered clock/calendars, and this entry is not required. Some of the first OMEGA/VLF systems also required that the pilot enter the magnetic variation for the area in use, however all of the newer units have eliminated this requirement—changes in magnetic variation are handled by the computer.)

The most important element in the initialization process for all OMEGA/VLF systems is accurate entry of initial position. Without accurate initial entry of present position, the computer will be unable to determine its initial lane when operated in the Absolute Mode, and will carry any error in the initial en-

try of position forward when operated in the Relative Mode.

Once all of this information has been entered or provided by the computer—date and time, magnetic

Fig. 10-8. This CDU uses color coding to categorize different types of data, and includes a built-in Course Deviation Indicator. Photo courtesy of Tracor Aerospace.

variation, initial position—the unit will go into a search-and-capture mode similar to that for LORAN-C, and begin tracking. At that point, the unit is ready for in-flight navigation.

RATE AIDING

OMEGA works with phase, which means it has to use the entire 10-second transmission of each station to plot an LOP, however, it takes two LOPs to determine a fix. During the 10 seconds it takes the computer to determine the second LOP, the aircraft will have moved and will no longer be where it was when the first LOP was determined—at a ground speed of 360 KTS, for instance, the aircraft will have moved one nautical mile during the 10 second computation cycle for the second LOP.

To correct for this movement, OMEGA/VLF uses a process called *rate aiding*. Rate aiding provides the computer with heading and true airspeed information, which it then uses to correct two or more LOPs to the equivalent time period. This is also called *advancing a position*—using speed and heading information, the computer advances the LOP to an estimate of where it would be 10 seconds later. In this way two LOPs taken 10 seconds apart can still be used to describe a fix. The more accurate the speed and heading information, the more accurately the LOP will be advanced, and the better the fix.

All OMEGA systems provide means for automatic entry of gyro-stabilized heading information; more sophisticated OMEGA systems also provide for automatic input of TAS. Other systems require manual entry of TAS, which is considerably simpler and less expensive, but which obviously increases the pilot's workload and decreases the overall accuracy: each time the TAS changes significantly (from climb to cruise, after an altitude change, or from cruise to descent), the new TAS must be re-entered.

In practice, rate aiding is generally not critical to overall system accuracy—very often, more than the minimum of two or three stations (Relative or Absolute Modes respectively) are available, and the computer can take advantage of these extra stations to plot extra LOPs, each of which reduces the interval between plots. In addition, the system software contains broad guidelines for TAS rate aiding, and

can extrapolate a fairly accurate heading using track. (This is an example of another kind of boot-strap operation, where a series of fixes is used to provide estimated heading, and that estimated heading then is used to aid in determining the next fix). Therefore, should either TAS or heading inputs fail, or be unavailable, the unit will still function adequately, but at reduced levels of accuracy. (In fact, the FAA will not certify an OMEGA system for IFR use unless it can function within limits without rate aiding.)

It is this rate-aiding feature that causes OMEGA to sometimes be described as a dead reckoning computer with 10 second position updates. To the extent OMEGA can function satisfactorily without rate aiding, but cannot function normally by dead reckoning alone, I think this description is not entirely accurate. (Most OMEGA systems do have the capability to temporarily navigate entirely via dead reckoning if OMEGA/VLF signals become insufficient for navigation, but OMEGA/VLF cannot function permanently by dead reckoning alone.) Probably the best general description of OMEGA is that it is a position fixing system that depends on dead reckoning for peak accuracy—a technical distinction perhaps, but it is important to know what part of the system is essential, and what part can be done without.

PROPAGATION ANOMALIES

OMEGA/VLF is subject to a variety of *propagation anomalies* (transmission inconsistencies), some of which are predictable, and some of which are not. All affect the accuracy and reliability of OMEGA/VLF navigational information. Diurnal effect (one of the predictable anomalies) has already been discussed. There are several others.

Sudden Ionospheric Disturbance

A *Sudden Ionospheric Disturbance*, or SID (not to be confused with a Standard Instrument Departure, also abbreviated SID), is a propagation disturbance caused by solar activity—sun spots. Solar activity causes radiation to be released, and the radiation creates charged particles, or ions. When these charged particles enter the ionosphere, they change its height, which affects the OMEGA/VLF signal phase in the same way diurnal effect does.

While the immediate effects of a SID normally last only for a few hours, long term effects caused by geomagnetic substorms often begin three to five days after a major SID, and last for another three to five days. Position errors of as much as 20 nautical miles can be caused by SIDs and by the geomagnetic substorms which follow a SID, although errors of five nautical miles or less are more common.

While the solar cycle of activity is well known and fairly predictable, having an average cycle period of 11.4 years, it is not predictable enough to include in the automatic compensation incorporated into every OMEGA/VLF system. Solar activity is predictable on a day-to-day basis however, and pilots using OMEGA/VLF are advised to request OMEGA NOTAMS as a part of their normal preflight briefing. During the three year peak period of the normal 11.4 year solar cycle, a major SID affecting OMEGA/VLF signal transmission quality can be expected, on average, once per day.

Polar Cap Anomalies

Some of the charged particles which follow a SID are deflected to the polar regions by the earth's magnetic field. These particles cause the ionosphere to drop in the affected regions (generally, above latitude 65 North and below latitude 65 South). This distortion is called a *Polar Cap Anomaly* (PCA), and errors of as much as 15 nautical miles can be expected (although errors from PCAs are normally less than six nautical miles). A PCA can last from a few hours to as long as 10 days. PCA warnings are also included in OMEGA NOTAMS.

If the OMEGA/VLF system in use includes a provision for manual de-selection of stations, then those stations that are being received along a polar path should be de-selected during a PCA. De-selection of stations increases position error, but less than the PCA distorted signal does. If manual de-selection is not available, the pilot should carefully monitor the OMEGA/VLF system for accuracy and reasonableness during the PCA warning period, and back-up the OMEGA/VLF position fixes with as many other systems as are available.

Modal Interference

Modal Interference is a propagation anomaly caused by the mixing of ground and sky waves. At relatively short ranges—200 to 500 nautical miles—modal interference is predictable, and the OMEGA/VLF system can compensate for these errors by automatically de-selecting stations in this range. At greater distances—up to 2,000 nautical miles—modal interference is primarily a function of diurnal shift during sunrise and sunset, and is called *extended range modal interference*. Extended range modal interference can occur at any time—it is essentially unpredictable and undetectable. It is potentially the largest and most serious source of error in the OMEGA system. Any large, unexplained discrepancy in position between the OMEGA system and any other system on board can usually be attributed to extended range modal interference.

Any time an OMEGA position report appears to be in error (compared to all other navigational sources), or is clearly unreasonable, the pilot should experiment by de-selecting specific stations and noting the new position reports that result. Modal interference will normally affect only one station at a time; by manually de-selecting stations, most combinations of stations will produce similar fixes (plus or minus two or three nautical miles). The combination of stations that includes the station experiencing modal interference, however, will differ significantly from the others, and once identified, should be manually de-selected.

Modal interference normally only lasts for a short period; after manually de-selecting a specific station, a re-check of other combinations should be made 30 minutes to 60 minutes after de-selection, and the system returned to automatic selection when modal interference is no longer a problem.

Wrong Way Propagation

Another error that is hard to detect is *Wrong Way Propagation*. Since OMEGA/VLF signals travel such long distances, the aircraft will sometimes receive not only the signal that travels along a direct path between transmitter and receiver, but also the signal that travels all the way around the earth in the other direction. This second signal will normally be much weaker than the direct signal, and the OMEGA/VLF system will automatically de-select it; however, in certain circumstances, for reasons that are not entirely understood and cannot be accounted for, the wrong-way transmission will sometimes be stronger than the direct transmission. Since this wrong-way signal has to travel much further than the right-way signal, it will arrive later, and a significant position error will result. Again, if the fix seems inconsistent or unreasonable, the solution is to manually de-select stations until the offending station has been identified.

Magnetic Deflection

Signals crossing the Magnetic Equator from east to west are deflected by the earth's magnetic field and are unreliable. For example, signals originating at the Liberia transmitter (refer back to Fig. 10-1) are unreliable along the east coast of the United States. Some systems include software to automatically de-select stations in which this anomaly can be expected; others require that charts be consulted for various locations, and specific stations manually de-selected. Errors of as much as six nautical miles can result from these unreliable signals.

Compensation

These various propagation anomalies affect different systems differently, depending on software and design choices. For instance, some systems emphasize software sophistication in an attempt to analyze and automatically de-select suspected stations. Others plot fixes using as many LOPs as possible and rely on sophisticated sampling techniques to arrive at a Most Probable Position. This later approach deliberately sacrifices a certain amount of ultimate accuracy in order to dilute the affects of any unpredictable and undetectable propagation anomalies.

Just as with LORAN-C, there is not a right or wrong way to determine position with OMEGA/VLF, and, as with so many things in aviation, the end product is inevitably the result of a series of compromises. The better the pilot understands his own system, the more he will be able to obtain from it, and the less likely he is to be misled by it.

OMEGA/VLF ACCURACY

Regardless of the techniques used, or the design choices made, and despite an occasional but rare gross error due to one propagation anomaly or another, OMEGA/VLF is still a very accurate and reliable long-range navigational system. In the Absolute Mode, most OMEGA/VLF systems are capable of position accuracies to within two nautical miles at all distances, in any location worldwide, regardless of the amount of elapsed time since initialization, with a 95 percent probability. In the Relative Mode, accuracy is directly related to the accuracy of the initial fix, with overall system accuracy degrading as function of time from that point on. The degree of degradation (drift) varies from manufacturer to manufacturer; 0.5 nautical miles per hour is a typical figure.

THE ROLE OF OMEGA/VLF

Other systems are more accurate than OMEGA/VLF (LORAN-C and DME), however their coverage is limited. Inertial navigation has complete worldwide coverage with fewer limitations than OMEGA/VLF, but inertial (INS) accuracy diminishes as a function of time, and in any case, INS is extremely expensive. OMEGA/VLF, operating in the Absolute Mode, is the only existing navigational system capable of providing consistent accuracies, worldwide, to within two to four nautical miles, regardless of the amount of time since initialization, and OMEGA/VLF, operating in the Relative Mode, is the only system that can provide worldwide inertial type navigation at one-half to one-fifth the cost of inertial navigation systems themselves. The existence of these unique characteristics is the reason OMEGA/VLF exists.

FUTURE PROSPECTS

The future for OMEGA/VLF is mixed. On the one hand, OMEGA/VLF in the Relative Mode does nearly everything the much more expensive, heavy, and complex inertial systems do, and OMEGA/VLF in the Absolute Mode can do what no other existing long-range system can do, and that is provide accurate, absolute position determination, worldwide, independent of elapsed time. On the other hand, OMEGA/VLF is a fully developed system with limited room for growth or improvement, while other systems, both existing and planned, have the potential for considerable expansion and continued development: INS accuracy is theoretically unlimited; a worldwide LORAN-C would almost certainly render OMEGA/VLF obsolete; the NAVSTAR/GPS satellite system promises to provide incredible accuracies worldwide at a very reasonable cost to the user.

OMEGA/VLF certainly has a place in the cockpits of many aircraft for the time being, both for individual pilots desiring a primary long range system with worldwide coverage, and for international operators looking for either a less expensive alternative to a second or as third INS, or as a complement to their existing INS systems. In the short run, OMEGA/VLF's place is secure. Whether this will always be the case or not remains to be seen—it is impossible to say with certainty what the future holds for OMEGA/VLF, since so much depends on what happens with LORAN-C and NAVSTAR/GPS in the next few years. This is a time of rapid change in air navigation, and the future for OMEGA/VLF is particularly unclear.

Chapter 11

Inertial Navigation Systems

Inertial navigation is a completely self-contained navigational system that is capable of providing Great Circle courses over random routes without the need to communicate in any way with any external sources of information. In this respect, it is unique among modern air navigation systems. Inertial navigation is also the heaviest, largest, most complex, and most expensive cockpit navigational system currently in use. Despite these drawbacks, the unique characteristics and capabilities of inertial navigation justify its purchase and installation for many operators.

While it is unlikely that the expense and size of Inertial Navigational Systems (INS) will ever be reduced to the point where INS is a practical reality for the average general aviation pilot, inertial systems are practical realities for those operations that regularly operate turbine equipment internationally, and to the extent INS is commonly considered to be the ultimate navigational system, this book would be incomplete if it were not covered. In any case, many pilots find INS to be a very interesting system, one

that brings together nearly all other aspects of navigation.

Inertial navigation, like so many things in aviation, was originally developed by the military. The military wanted a system that was completely self-contained—a system that could function completely independently from any external visual or electronic reference. Initially intended for submarine navigation, INS was later extended to surface ships, and eventually to aircraft.

The strategic and tactical advantages of a self-contained navigational system are very significant: an independent system can function despite the total destruction of all land-based facilities (such as TACAN and OMEGA), it cannot be jammed, and it cannot be fed reasonable seeming but erroneous information. An independent system is capable of operation worldwide, and its accuracy is limited not by ground facilities or limitations inherent in the design of the system, but only by technology: there are no theoretical limits to inertial accuracy, only practical and technological considerations. INS is the only navigational system where it can truly be said that the operator can buy the accuracy he requires—the more sophisticated (and therefore the more expensive) the system, the greater the accuracy. All of these factors make INS the navigational system of choice for the military.

INS is also the navigational system of choice for those operators who frequently travel internationally and over water. No other long-range system provides the consistency, reliability, and freedom from limitations that INS does. For these reasons, INS is often the primary and sometimes the only long range navigational system in use on many internationally operated aircraft.

As in other chapters, this chapter is not meant to be a substitute for the training and instruction necessary to safely operate an inertial navigation system; this chapter is meant to provide a general description of how inertial systems work, what they look like, how they are initialized, how they differ from OMEGA/VLF and LORAN-C systems, how they can be used most effectively, and the essential differences between dual and triple installations of INS units. This chapter is meant, in other words, to be an introduction to INS—a bridge between the little that has been written about INS for general aviation in the past, and the detailed technical manuals and training provided by the manufacturers.

GENERAL DESCRIPTION

In theory, nothing could be simpler than INS, and in practice, almost nothing in air navigation is more complex. We will make no attempt to describe in detail how an inertial system actually works, or describe all of the various ways it corrects for a variety of complications, but we will look at inertial systems in general terms. (Those with a solid background in science and engineering may want to write to DELCO [Appendix A] for a copy of their excellent book on basic inertial system theory called [*Carousel IV & IV-A, Inertial Navigation Systems: System Technical Description.*]

INS has been described as a very accurate dead reckoning system, and to the extent INS starts from a known point, just as dead reckoning does, and bases its subsequent estimates of position on speed, direc-

tion, and time from that point, this is an accurate statement. It is, however, somewhat of an oversimplification. INS works directly with *acceleration*—changes in aircraft speed or direction—not speed itself. Acceleration (and deceleration—negative acceleration) are used to obtain changes in direction and speed, which in turn are used to estimate new positions based on projections from the original position; this is at least one step removed from dead reckoning per se, but describing it as such is an effective way of summarizing how INS works.

In theory, INS works in a very simple way. It says, "If I know where I am now, and if I keep track of each and every movement in all directions from this point on, then I will always know where I am after that point." To detect movement, INS uses accelerometers—three of them, one aligned North and South, one aligned East and West, and one aligned vertically—all mounted on a stable platform. The platform is stabilized with gyros—three of them—one for each of the three primary control axes (pitch, roll, and yaw).

An accelerometer is a relatively simple device that detects acceleration, or changes in velocity—a pendulum would work, but sensitive devices employing sliding shuttles with essentially frictionless bearings are more often used. These devices can detect accelerations to thousandths of a G force—far more sensitive than the human body can detect, for example.

Further improvements in accelerometers are always possible, which would result in further improvements to INS position estimates, but the gains at this point would be slight. Surprisingly, despite the fact that the entire system is predicated on measurements of acceleration, this is not the most critical component—platform stability is much more critical to overall system accuracy than accelerometer sensitivity, and most of the improvements in INS operation over the years have come about through advancements in the gyro technology necessary to stabilize the platform, not in accelerometer technology.

Gyros are critical for two reasons: gyros keep the accelerometers aligned horizontally with the surface of the earth despite changes in aircraft attitude, and gyros provide the sensing to keep the stable platform aligned vertically with the earth as the aircraft changes location, and as the earth itself rotates. These factors may require a little more explanation.

The first problem is easy to understand: If the inertial system were simply bolted to the floor of the aircraft and the aircraft were to roll into a turn, or change pitch, or yaw, the applicable accelerometer would no longer be aligned along its original axis, but would be tipped, and the accelerometer would indicate a false acceleration. The solution is to mount all three accelerometers on a gyro stabilized platform that maintains its vertical alignment, regardless of aircraft attitude.

This solves one problem, but causes another, one that is not so obvious: a freely mounted, gyro-stabilized platform will attempt to maintain itself in exactly the same plane at all times, even if the aircraft has moved to another position on the earth where the vertical orientation is not the same as the previous position. For instance, a stable platform that is aligned exactly parallel with the surface of the earth at JFK airport (New York), that is then flown to Buenos Aires, Argentina, would be nearly upside down by the time it got to Argentina. An aircraft traveling from the Northern Hemisphere to the Southern Hemisphere executes, in effect, part of an outside loop as it parallels the surface of the earth from North to South, but the gyro, in maintaining exactly its original orientation, would not execute this loop, and would appear, to the person looking at it in Argentina, to be upside down. Actually, it is the aircraft that is upside down (just as people in Antarctica appear to be standing upside down to someone on the North Pole). The same thing happens with east-west movement—people in China appear to be upside down compared to people in New York (as do we to them), and the uncorrected stable platform that originated in New York would appear to be upside down to people in China. The gyro stabilized platform must, therefore, be corrected so that it will remain oriented to local vertical and not original vertical.

To further muddy the waters, a gyro that is aligned parallel to the surface will appear to tilt over time, even if the aircraft does not move. This is be-cause the earth rotates at a rate of 15 degrees per hour (360 degrees divided by 24), and a free gyro cannot tell the difference between changes in the actual aircraft attitude and changes caused by the earth's rotation—an aircraft, parked on the ground at the Equator and pointed due east (for example), will appear, to the gyro, to be slowly diving as the earth rotates about its vertical axis.

Both of these errors (the gyro is only doing what it is supposed to do, but in so doing it introduces what appear to be errors) must be accounted for, and the process of correcting for all extraneous gyro factors is called torquing. Torquing is engineering jargon for a feedback process whose purpose is to keep the stable platform level with respect to local vertical.

We will not go into the mechanical details of this torquing process, but the mere existence of torquing is itself revealing: INS may appear to be a simple system, but it is not. It also helps explain why inertial systems are so complicated, heavy, and expensive, and why inertial accuracy is a direct function of technology and nothing else: the more sensitive the accelerometers, the more stable the gyros, and the more sophisticated the torquing software and hardware, the more accurate the system.

In state-of-the-art inertial systems (1986), mechanical gyros have been replaced by ring lasers—a series of lasers whose beams are aligned in the same plane and form a ring. Interference patterns created by shifts in the relationship of the laser beams as the plane tilts are used to indicate changes in the plane. Ring lasers are more accurate and more reliable than mechanical gyros, they consume less power and generate less heat, and they are lighter. Since they are not true gyros, they do not have to be freely mounted but can be strapped down, which solves some problems and creates others (all of which are corrected digitally in the computer, rather than physically). Ring lasers will no doubt completely replace mechanical gyros eventually, and strap-down systems will become the norm.

Improvements in inertial technology have resulted in increasingly more accurate INS systems. The first inertial systems were accurate to about four nautical miles per hour (NMPH)—which means that the system could be expected to indicate actual position to within four nautical miles after the first hour, an additional 4 nautical miles after the second hour, and so on. Present airline level INSs are capable of accuracies of about 1.7 NMPH, and the military has inertial systems capable of accuracies as low as 0.2 NMPH. (No doubt the military also has classified systems capable of even greater accuracies.)

INERTIAL REFERENCE SYSTEMS

The gyros used to stabilize the platform on which the accelerometers are mounted are much more accurate than the gyros in even the very best attitude indicators and directional gyros. It would therefore be a waste to use them only for inertial system stabilization. All modern inertial systems are able to use the gyros in the inertial system (or systems) to supply attitude and heading information to the cockpit indicators (normally to a Flight Director and Horizontal Situation Indicator), to stabilize the weather radar antenna, and to control the autopilot. This part of the INS is called the Inertial Reference System (IRS), as distinct from the Inertial Navigation System itself. Inertial Reference Systems significantly improve the performance of these components. If the IRS fails, or is manually de-selected, the normal attitude and heading gyros (which are still installed) automatically take over.

A mode selector panel is normally provided with all inertial systems, allowing for selection of either the combined navigation and reference functions, or the reference function alone. A mode selector panel

Fig. 11-1. The accelerometers, gyros, and computer for the Delco Carousel Six INS are contained in the large ARINC Sensor Unit box designed for remote mounting. The mode selection panel is normally console mounted and includes a NAV READY light and an emergency battery low level warning light. Photo courtesy of Delco Systems Operations.

for the Delco Carousel INS, an older but widely used and highly respected inertial system typical of many inertial systems, is shown in Fig. 11-1, along with the Sensor Unit. (All of the gyros and accelerometers are contained in the Sensor Unit, as well as the computer.) The mode selector panel has a switch marked *OFF/STBY/ALIGN/NAV/ATT*, a READY NAV light, and a battery low level light (BAT). The battery unit (not shown here) is used primarily for emergency power.

The NAV position is the normal navigation and inertial reference position. The ATT position (for attitude) is the position for inertial reference only. If the nav portion of the INS fails with the selector in the NAV position, the unit will automatically revert to inertial reference without having to select ATT; the ATT position is provided in order to manually disable the nav function while retaining the attitude function. The pilot might select this position when operating on short, domestic flights along airways where long range, area navigation information was not required, but where he wished to retain the inertial attitude and heading information.

CONTROL-DISPLAY UNITS

There are two basic types of INS Control-Display Units in general use: the airline/military type incorporating the traditional two window lat/long readout, and the cathode ray tube (CRT) type found on many corporate turbine level units. (These latter types are also very similar to the CRT CDUs illustrated in the chapters on LORAN-C and OMEGA/VLF.) A traditional CDU is illustrated in Fig. 11-2, along with the remaining components for a complete INS system. With the cover off the Inertial Navigation Unit, the gyro stabilized platform can be seen in the back half of that box. The CDU for the Delco Carousel SIX INS system, a CRT type CDU, is illustrated in Fig. 11-3.

There are advantages to each type of display, although the CRT is generally preferable, and eventu-

Fig. 11-2. A complete Inertial Navigation System, showing (from left to right) the standby Battery Unit, the Inertial Navigation Unit, the Mode Select panel, and a traditional, two window Control-Display Unit. Photo courtesy of Delco Systems Operations.

Fig. 11-3. The Control-Display Unit for a modern INS with a Cathode Ray Tube type of data display. The CRT type CDU can display large amounts of data, including complete flight plans. Photo courtesy of Delco Systems Operations.

ally all INS CDUs will probably have CRT displays. The advantage to the traditional type is simplicity—only one bit of information can be displayed at a time, which eliminates confusion. The numbers are also a little bigger, brighter and easier to read, and the circular selector lists all the possible read-outs in a straightforward manner. The disadvantages are obvious: only limited information can be displayed at any one time, and the ability to store and retrieve complete flight planning information over various routes is not possible. With this type of CDU, flight plans must be manually entered using departure, waypoint, and destination lat/long coordinates each time a particular trip is flown.

The CRT type CDU solves these problems at the expense of a little additional complexity. The INS system associated with the CDU shown in Fig. 11-3 can store flight plans for up to 20 routes with permanent storage of up to 100 waypoints (i.e., an average of 5 waypoints per route, if all 20 routes are used). The ability to store complete flight plans is particularly valuable for corporate operations that regularly operate from a given base to several different destinations. It is less valuable at the present time in airline operations: in airline operations, a particular aircraft might find itself anywhere on any given day throughout an

extensive domestic and international route system. To cover every possible route within that airline's system would require hundreds and perhaps thousands of stored flight plans, and that is beyond the capability of present INSs.

The main advantage of the CRT-type display is that it can display several lines of data at once. Rather than select individual bits of data with a selector knob (as the traditional type does), the CRT displays as much data as possible at all times, and allows the pilot to see more data by changing ''pages'' (a term borrowed from computer technology, which simply means to change from one list of data to another). Pages are categorized by type, i.e., POS for Position, PLAN for Flight Plans, and so on—see the key markings in Fig. 11-3. Repeated pressing of any given category key changes from one page in that category to the next.

The operation of a CRT display is a little more complicated than the operation of a simple two window type of display, and it demands a certain level of proficiency to be able to take full advantage of its capabilities. Neither type of display is especially difficult to use though—airborne INSs are highly automated systems, and were designed from the beginning to be easy to use in the crowded cockpits of high speed, long range aircraft.

INITIALIZATION

INS initialization is a lengthy and complicated process internally, but a relatively simple one for the pilot. The entire process of initialization is automatic after pilot entry of date, time, and ramp coordinates (position on the field). In fact, most INSs have internal clock/calendars, so in most cases the pilot only needs to enter the ramp coordinates, and then verify that the date and time are correct. (The military has INSs which can be initialized in the air, but all civilian INSs must be initialized on the ground with the aircraft in a stationary, known position.)

The entire process, from system turn-on to nav ready takes from two and one-half to 45 minutes, depending on the ambient temperature (cold gyros take longer to spin up than warm gyros do), gyro type (lasers are faster than mechanical; early gyros were quite slow), and latitude (the higher the latitude, the longer it takes the gyros to sense earth rotation and find True North).

The various steps and times for the DELCO Carousel system are as follows:

1) *Standby Mode.* Internal heaters bring the system up to temperature; gimbals are caged in the pitch and roll axes, and gyros spin-up. Time: as much as seven minutes.

2) *Align Mode 8:* Gimbals are uncaged, coarse leveling of the stable platform is accomplished and gimbals are aligned with local vertical. Time: as much as one minute.

3) *Align Mode 7:* Coarse azimuth alignment is accomplished by direct sensing of earth rotation to locate True North. Earth rotation determines rough estimate of latitude, which is compared to entered ramp latitude coordinate to check for reasonableness. Time: about four minutes.

4) *Align Mode 6:* Fine alignment of True North, using ramp coordinates for increased precision, and calibration of gyros and accelerometers to account for instrument and installation errors. Fine alignment may take as long as six and one-half minutes.

5) *Align Mode 5:* Minimum alignment and calibration standard achieved for en route navigation. System is ready to use, READY NAV light illuminated. Mode selector may be advanced to NAV.

6) *Align Modes 4-0:* If time allows, the system will continue to refine the vertical and horizontal alignments to a maximum of Align Mode 0. This would indicate maximum system capability, with resulting incremental increases in system performance.

From the pilot's point of view, initialization is simple: select Standby and then Align mode, enter ramp coordinates, check date and time (and correct as necessary), then wait for the Nav Ready light to come on and select Nav. The aircraft can then be moved and the INS will begin sensing and correlating all aircraft movements, no matter how small or large, slow or fast, and regardless of aircraft attitude or location. (The only exception to this rule is that certain systems are unable to deal with the extreme longitudinal convergence that occurs near the poles, and cannot be used reliably above 80 degrees North latitude and below 80 degrees South latitude. This

is a very minor restriction for most operators, since few civilian aircraft operate this close to either pole, and in any case most INSs now use a type of orientation system called a *free azimuth system* that gets around even this restriction.)

EN ROUTE DATA

With acceleration information about three axes, each perpendicular to each other (an orthogonal relationship), aircraft movement, and therefore position, can be continuously determined in three dimensions. The CDU, however, reports in only two dimensions—latitude and longitude; altitude—the third dimension—is used only for internal computations.

Since the INS is capable of sensing all movement, and since it is always knows where True North is and therefore always knows what the aircraft heading is, it is a simple matter for the computer to determine such additional information as:

Track: a series of fixes defining a line, or track.
Ground Speed: time between two points of known distance, converted to an hourly rate.
Drift Angle: the difference between aircraft heading and track.
Cross Track Error and Cross Track Distance: comparing Actual Track to Desired Track, both in angle and in distance.
Distance-To-Go and Time-To-Go: present position compared to the selected waypoint, and distance divided by ground speed.

All INSs normally provide for automatic input of fully corrected TAS through a separate system called an *Air Data System.*. The INS computer uses this fully corrected TAS to determine wind direction and velocity (just as wind direction and velocity can be derived with an E-6B by working a wind triangle problem backwards).

Wind direction and velocity is the only bit of derived navigational data that requires an outside source of information, and that is contained within the aircraft. In the event TAS input fails, the INS will continue to function exactly as before the failure, however wind direction and velocity will not be directly available—a relatively minor and normally inconsequential consideration.

DRIFT

Inertial systems are, by definition, relative systems—every movement recorded has meaning only to the extent it is related to another position. Therefore, any error in the initial entry of position will be carried as a constant value for the course of the flight (or until corrected), and all subsequent errors are cumulative, increasing as a direct function of time. The sum of these errors is called drift (presumably because this gradual accumulation of errors causes the aircraft to appear to drift from the true position). Drift is normally measured in nautical miles per hour; 2 NMPH is considered to be an acceptable, industry standard.

Two NMPH may sound like a lot of drift for a system that is supposed to be the ultimate in navigational sophistication, but it is important to put that drift rate in context: Even though some aircraft are capable of 10 and 12 hour flights, all routes, regardless of the length, have portions at the beginning and end, and sometimes in between, that are over land and within range of supplemental navigational aids. Even the longest over-water routes are dependent exclusively on INS for only a few hours—generally no more than about five hours for turbine aircraft. For instance, turbofan aircraft operating at normal cruising altitudes and speeds between San Francisco (SFO) and Hawaii (HNL) will be out of VOR/DME range for about four hours and 45 minutes with average winds, and this leg is one of the longest overwater routes in the world without any intermediate, ground based navigational aids. (Slower aircraft with extended range tanks will be out of VOR/DME range for longer periods than this on this route, but slower aircraft are seldom if ever equipped with INS—for slower aircraft, non-drifting navigational systems such as LORAN-C or OMEGA/VLF in the Absolute Mode are preferable to INS.)

With a drift rate of two NMPH, the total cumulative drift will be about 10 nautical miles on this route between the mainland and Hawaii. Normal separation standards on this route are 50 nautical miles and 1,000 feet: even if two conflicting aircraft were to drift 10 nautical miles directly toward each other, they would still be separated by 30 nautical miles and 1,000 feet. Thus, in actual practice, INS drift rates of two NMPH are more than adequate for safe aircraft separation and navigation.

UPDATING

Drift can be corrected by updating. This means taking advantage of other sources of navigational information to provide a new starting point, cancelling all or much of the accumulated error in the INS up to that point. Updating can take the form of re-entry of position while passing over a known fix (VOR or NDB station passage), passage over a known visual check point (the center of an airport, for instance), a VOR/DME fix, dual DME fixes, and fixes from supplemental long range information such as that provided by LORAN-C or OMEGA/VLF (Absolute Mode only—Relative Mode is also subject to drift).

While these updated entries cannot be expected to be as accurate as the initial entry of position, to the extent they are more accurate than the accumulated error up to that point they will prove beneficial. After two or three hours of drift, fixes based on other electronic estimates of position will generally be more accurate than the INS at that point, and should be used.

The specific decision as to whether to use an updated position fix of somewhat uncertain accuracy, or rely on the known drift rate of the INS, is a matter of pilot experience and judgement, and is highly dependent on the specific situation and the characteristics of the individual systems being used. The higher the quality of the supplemental fixes, the greater the drift rate of the INS, and the longer the time since the last update, the more the INS will benefit from updating. For this reason, it is very important to understand the workings of all navigational systems when using INS; without a knowledge of relative system characteristics and accuracies, intelligent updating of the INS cannot be done.

DUAL INSTALLATIONS

In aviation, redundancy is a fundamental requirement for any critical system, and for long range navigation, where other sources of navigational guidance are not available, particularly when aircraft separation is also predicated on accurate navigation, dual long-range systems are essential (and are often required, as we will see when we get to over-water navigation—Chapter 13). Dual systems not only provide the necessary redundancy, but with two systems, each system can be used as a check on the other.

Most INSs are programmed to detect certain input errors (such as an impossible coordinate—North 100), but by and large, if the pilot enters incorrect information into the INS (the computer), the INS will report incorrect information back (garbage in, garbage out). The best way to detect these kinds of errors (reversing coordinates, entering North for South, punching-in incorrect numbers, etc.), is to enter all route information into each system separately, preferably by two different crew members, and then compare the results: each system should show the same desired track, the same distance to destination, and the same present position. (Another technique is to have one crew member read the coordinates while the other enters them, then reverse roles for the second system.) If a discrepancy is noted between the two systems, then one of the two systems is either incorrectly programmed, or is malfunctioning. (The latter possibility is much less likely than the former—when inertial systems malfunction, they almost always display either a clear warning indication, or else they simply do not work, displaying instead a blank screen, or all eights in the data windows.)

Many inertial systems have provision for remote or automatic entry of initial and route information into second and third systems—once one system is loaded, the second and third systems can be automatically loaded by punching Remote. The obvious disadvantage to this convenient feature is that if a mistake is made in programming the first system, that mistake will be transferred to each of the other systems, and one of the best ways of catching it—comparing one system against another—will have been lost.

Assuming both systems agree and seem to be producing reasonable results, both systems should continue to agree, within the expected accuracy limits for each INS, for the duration of the trip. If one disagrees with the other by a significant and unexpected amount, it is important not to jump to conclusions—the one which appears to be incorrect may, in fact, be the correct one. With only two systems, and without any additional supporting information, it is impossible to know for sure which one is the incorrect one.

If the supporting information is available—OMEGA/VLF, VOR, DME, NDB, perhaps even a radar ground-mapping fix (see Chapter 15)—then it may be possible to determine which system is correct (or most correct) and which is not. In the absence of any supporting information, dead reckoning can sometimes be used in an attempt to identify the correct system. Unfortunately, unless the error is a very large one, dead reckoning will seldom be accurate enough to identify an incorrect INS.

There is a temptation in these kinds of situations to average the two positions in an attempt to dilute the error. This is not always the best choice. Averaging sacrifices the correct system in order to minimize the error caused by the incorrect system.

A better way to reconcile the discrepancy is to attempt to predict which system is the correct one, and there is only one way to do that with any certainty and that is to keep complete records on the performance—the drift—of each system. With accurate record keeping, it should be clear over time which is the more accurate system. In many cases there will be additional supporting evidence, either external (other navigational sources), or internal (warning indications), and this difficult choice will not have to be made. However, in the absence of any additional information, the system with the record of maximum accuracy should be relied on and the other disregarded. If the correct system cannot be identified with any assurance, then the two should be averaged.

TRIPLE INSTALLATIONS

The ultimate navigational system is a triple installation of inertial systems—three entirely independent, but interconnected INSs. In addition to providing double redundancy and a second check on route entry and initialization, triple systems have the advantage of being able to provide key supporting information when two systems disagree. In addition, by a process called *triple mixing*, three systems can be used to provide accuracies that are nearly double that of systems operated independently or in parallel.

Voting

Triple INSs can be used to identify the weakest system in a group of three by *voting*: a vote is taken by all three systems, and the two that are closest together win and the third is rejected. It is always possible of course, that the system with the greatest difference in position is more correct than the others, but it is much more likely that the two that more or less agree are correct, and that the odd system is the incorrect one. By adding a third system, there is always a way to identify the system with the greatest probable error and eliminate it.

Triple systems also provide the only way to catch what is the most illusive and dangerous type of malfunction, and that is the malfunction that occurs en route, out of range of other navaids, that appears reasonable and normal, but is not. With only one system, this error cannot be detected at all. With two systems, the error can be detected, but little can be done about it except to assume that the unit with the best record is the correct one. However, with three systems, if one system is in error, two will tend to agree and one will clearly disagree. The odd system—the rogue as it is sometimes called—is thrown out. In this way, malfunctions that may not be obvious can be detected.

Triple Mixing

Some systems are capable of mixing position estimates from three independent systems to arrive at a single, optimum estimate of position. This is called triple mixing, and it is a much more complicated task than either voting or averaging. (Fortunately, the computer does all the work.) The results are remarkable though: An individual Delco Carousel Six INS (illustrated in Fig. 11-2), has an accuracy of 1.7 NMPH with a 95% probability; accuracy increases to 1.0 NMPH, again with a 95% probability, with triple mixing—nearly double.

TRIPLE MIXING VERSES OMEGA UPDATING

There is no operational advantage in integrating OMEGA/VLF with INS if the OMEGA/VLF system is limited to the Relative Mode—both systems will drift at about the same rate, and neither is capable of reliably updating the other. However, OMEGA/VLF with Absolute Mode capability can be used to reliably update INS. OMEGA/VLF accuracy tends to be fairly constant (about two nautical miles) at all times, which means that the crossover point—the point where the OMEGA/VLF fix becomes more accurate than the INS fix—will occur after about 1 hour and 10 minutes, assuming a 1.7 NMPH drift rate. INS and OMEGA thus appear to be ideally complementary systems, INS providing a high degree of reliability and a stable inertial reference source, while the OMEGA/VLF system bounds, or limits, overall system inaccuracy to a maximum of two nautical miles with continuous updating. In fact, INS and OMEGA/VLF are frequently installed together for just these reasons.

Since at least two long range systems are required for high level North Atlantic crossings, most operators install three systems in order to have a spare for dispatch reliability. This leaves the person responsible for equipping and purchasing long range, internationally capable aircraft with an interesting and difficult choice: should he install three INSs, in order to be able to triple mix (assuming cost is not a major factor), or should he install dual INSs with an OMEGA/VLF system for updating? (A third choice is a single INS with dual OMEGA/VLFs, but this choice offers no advantage over either of the first two choices except reduced cost, and we are assuming that cost is not a factor here.) Triple INS will provide predictable, highly reliable, and reasonably accurate position fixes (one NMPH), whereas OMEGA/VLF is capable of providing position updates that are normally accurate to within 2 nautical miles for any length of time, but that are subject to various propagation anomalies and are therefore not quite as reliable and predictable as INS. Which, in other words, is more important: consistency and reliability, with some possible sacrifice in total accuracy (triple INS), or maximum average accuracy over time, but with a slightly greater degree of uncertainty (dual INS with OMEGA)?

There is no right or wrong to this question, and

neither the FAA nor ICAO recommends or requires one approach over the other—either is acceptable for international and long range navigation requirements. (In fact, triple systems are not required at all for overwater operation, which makes the lowest cost triple installation—single INS, dual OMEGA/VLF—still quite a bit better than the minimum.) Each increase in INS accuracy and each improvement in OMEGA/VLF changes the equation. At the present time, valid arguments can be made for either choice, and to the extent cost is a factor, substituting an OMEGA/VLF for an INS certainly makes sense, but the situation bears watching, especially as regards emerging technologies (i.e., satellite navigation).

INS AND GENERAL AVIATION

INS's position at the upper end of the general aviation market is well established; the prospects for INS at other levels of general aviation are somewhat limited. Even if system weight and size can be reduced (the lightest unit currently available weighs 48 pounds, split between one large and one medium size ARINC unit), and even if volume production were to reduce costs somewhat, there are inherent limits to what can be done: INS accuracy is directly related to technical sophistication, and any attempt to reduce size, weight, and cost simultaneously inevitably compromises accuracy. An inexpensive INS could be produced, but it would probably be only about as accurate as basic dead reckoning and would not be worth the cost, however reasonable it might be. An affordable INS is probably something that will never exist.

It is possible the newer systems of navigation will render INS obsolete for all operators, but my guess is that INS is too good a system and too much money and experience has gone into it for it to ever disappear completely, regardless of what happens with LORAN-C and NAVSTAR. I would also guess that INS will probably always be a high-ticket item though, found only in those operations that demand and require the optimum in independent, long-range navigation, regardless of cost.

INS is still a very worthwhile system for all pilots to be familiar with. The principles behind INS are fundamental to all aspects of air navigation, and the pilot who understands INS understands air navigation.

Chapter 12

Satellite Navigation

The ability to navigate via satellite guidance promises to be the single most significant advance in air navigation since the development of the VOR/DME system. While it will still be several years before the proposed NAVSTAR/GPS system of satellite navigation is fully operational (not until at least 1989), parts of the system are already in place, and it is not too soon to begin planning for the day when the system will be complete. Ultimately, satellite navigation may very well replace all other ground-based systems as a primary source of navigational information, including VOR/DME, NDB, OMEGA/VLF, and possibly even LORAN-C. (It is unlikely that all of these systems will completely disappear, but some will, and the rest will almost certainly be reduced to supporting roles once satellite navigation has been fully integrated into the National Airspace System.)

The NAVSTAR/GPS satellite navigation system is significantly different from all other types of navigational systems for several reasons:

(1) NAVSTAR/GPS will be able to report position anywhere in the world in three dimensions at two levels of accuracy. In the Precision (military) Mode, position will be accurate to within 16 meters, both laterally and vertically; in the Coarse/Acquisition (civilian) Mode, three-dimensional position will be accurate to within 100 meters.

(2) NAVSTAR/GPS will function virtually free of all environmental limitations, which means that it will work equally well at all times of the day or night, during all seasons, in all weather, and without solar disturbances.

(3) NAVSTAR/GPS will be a relatively inexpensive system from the user's point of view. While the satellites themselves are enormously expensive, it is anticipated that satellite navigation units will be no more expensive than comparable VOR/DME units.

(4) NAVSTAR/GPS is designed to be expandable to include CAT III precision-approach guidance, data transfer, position reporting and flight following, and aircraft conflict reporting and resolution.

These are the general outlines of the satellite system of navigation. The specifics—implementation schedule, operating modes, accuracy under actual conditions, cost, civilian access, data transfer functions, integration with other systems, and the role of altitude reporting—are still under development, and the exact outline of what will and will not be available, how the system will work in practice, and even when it will be available have not been completely resolved.

Part of the reason for the lack of more specific information at this point is that satellite navigation, like VLF, is primarily a military system. Its development and installation is being paid for by the Department of Defense (DOD), and the military role has the first priority. Unlike VLF, however, a civilian navigational component has been designed into the system, and use of the satellite system for air and marine navigation, as well as for other civilian uses (such as for truck, rail, and other fleet operations), has been mandated. Civilian users are assured access to the system, but the exact form that that access will take remains to be seen.

Satellite navigation is, of course, a generic term—there can be many different types of satellite navigational systems, and in fact, a specialized satellite navigational system, called Transit, already exists. (Transit has practical value only for ships and surveyers, since fixes can be obtained with this system only once every 90 minutes.) The specific satellite navigation system that we are presently concerned with is called the *Global Positioning System* (GPS), or NAVSTAR—both names are used interchangeably at this point; in this book, we have elected to combine the two names and call this system NAVSTAR/GPS.

The remainder of this chapter looks at how the NAVSTAR/GPS system is put together, how it works, what the current schedule is for implementation, what its impact on the rest of air navigation might be, and the best guess as to the outlook for general aviation regarding NAVSTAR/GPS. This discussion will, of necessity, be general in nature; even after the system is fully operational, it will no doubt be several years before its full impact has been absorbed. This chapter is intended, then, to be an introduction to satellite navigation for the purpose of familiarization and planning, and is subject to revi-

sion as experience with the system is obtained.

GENERAL SYSTEM DESCRIPTION

The NAVSTAR/GPS system—as presently planned—will consist of 18 satellites plus 3 spares, evenly spaced around the globe, all orbiting at an altitude of 10,900 nautical miles. (Presumably, collision avoidance has been anticipated.) A relatively high orbit was chosen for two reasons: a high orbit makes satellites difficult to intercept and destroy and fewer satellites are required for continuous, worldwide coverage. At this altitude, each satellite will orbit the earth twice per day (plus or minus about four minutes), and the spacing of the orbits will insure uniform coverage over all areas of the earth at all times, although the degree of uniformity will vary somewhat during these 12 hour orbit periods. The variation will, however, be predictable, and navigational coverage, within system limits, will still be available at all times.

All 21 satellites will be controlled by a Control Segment on the ground consisting of four Monitor Stations, an Upload Station (to communicate with the satellites), and a Master Control Station. The Control Segment monitors total system performance, corrects satellite position, and re-calibrates the onboard atomic time standards as necessary (normally once every day).

This system of orbits and control ensures that at least four satellites will be in view at all times, anywhere on the surface of the earth (the minimum number for a fix), and it also ensures that the exact position of each satellite will be known at all times with a very high degree of accuracy and that each satellite will have an extremely precise time standard. Each of these factors is a necessary prerequisite for accurate, three-dimensional position determination.

THEORY OF OPERATION

NAVSTAR/GPS works very much like DME, with one important difference: DME measures the elapsed round-trip time for a pulse of energy to be transmitted to a DME ground facility and be returned, and converts the resulting time into distance by dividing the time in half and multiplying it by the speed of light. (Refer back to Chapter 3, Distance Measuring Equipment, for a complete discussion of DME theory.) NAVSTAR/GPS, on the other hand, measures directly the amount of time it takes a signal to travel from satellite to receiver, and that time is then converted into distance from the satellite by multiplying it by the speed of light. Distance from the satellite results in a Line of Position (similar to a DME arc), and just as DME determines a fix by the intersection of two DME arcs (rho-rho navigation), NAVSTAR/GPS uses three distance measurements to define an intersection in a three-dimensional framework—latitude, longitude, and altitude.

The major difference between the way DME works and the way NAVSTAR/GPS works is that DME is an active system, while NAVSTAR/GPS is a passive system. DME must be initiated by a signal from the user—an interrogation. NAVSTAR/GPS works all the time and does not require any interrogation by the user. (The military wanted a passive system—interrogations can be intercepted, jammed, modified, and targeted.) Active systems can use simple and inexpensive quartz oscillators to time the interval between the start of an interrogation and the receipt of a return pulse, but passive systems normally cannot make accurate, direct measurements of elapsed time unless both units have been exactly synchronized, which requires an extremely stable clock at each end. "Extremely stable" in this case means stable to within 10^{-13} seconds per second—about .003 seconds per thousand years. Only atomic clocks are capable of this kind of stability; they are, however, incredibly expensive.

One way around this problem is to work with time differences rather than with time directly. (This is the approach taken by LORAN-C and OMEGA/VLF.) NAVSTAR/GPS gets around the problem in another way: NAVSTAR/GPS accepts the fact that a quartz oscillator at the receiving end (the aircraft end) will be somewhat inaccurate, and uses an LOP from a fourth satellite to correct for that error.

The satellites themselves do have atomic clocks on board; in fact, for accuracy and redundancy, each satellite will have not one but four atomic clocks, each of the most advanced type, called hydrogen masers. The installation of four hydrogen masers on each satellite is one of the many reasons each of these satellites is so expensive.

This general description of satellite system operation may be sufficient for many pilots at this point. The next few sections explain the actual working of the system in a little more detail. Those desiring even more detailed information can write to the Institute of Navigation (Appendix A) for a copy of their booklet entitled *Global Positioning System* ($10.00). There have been a few changes since this document was printed (a reduction from 24 to 21 satellites, for instance), but it is still the single best source of technical information on NAVSTAR/GPS.

Signal Structure

The NAVSTAR/GPS satellite transmits on two frequencies in the Ultra High Frequency band and in two modes. The first mode is called the P, or Precision Mode, and the second is the C/A, or Coarse/Acquisition Mode. The P Mode is for military use only and will not be available to the public (or to other, unauthorized military users.) The C/A Mode is the civilian mode.

In each mode, a pattern of signals is transmitted that appears to be random but that actually is not—the pattern is generated in a somewhat predictable and repeatable pattern by a computer and is called a *pseudo random code*. The P code is transmitted at a rate of 10,230,000 bits per second and takes a full week to repeat itself. (Multiplying 10,230,000 by the number of seconds in a week will give you some idea of the size and complexity of this code.) The C/A code is transmitted 10 times slower (1,023,000 bits per second), and repeats every millisecond. The C/A code is therefore much simpler, 1,023 bits in all, and can be deciphered fairly quickly even by unauthorized users. These differences in speed and size determine the degree of security, accuracy, and access time (initial computation time) between the two modes.

The NAVSTAR/GPS receiver has the same program for generating the P and C/A pseudo random codes as the satellite does; by matching the two patterns, the satellite and receiver can be synchronized, which is the first step in obtaining an LOP. In practice, the long cycle period of the pseudo random P code (seven days) is not only very difficult for an unauthorized user to match, but it is also very difficult for an authorized user to match, since the computer must search and compare codes many hundreds of thousands of times to acquire, or find the right spot in the pattern, even if it knows the code. The C/A code, on the other hand, repeats every thousandth of a second, and can be acquired quite quickly—seconds instead of hours. Therefore, the C/A code is always used for initial acquisition, and then, by a complicated process involving a hand-over signal, those authorized to use the P code switch over to that code for precision position information. The civilian or unauthorized user uses the C/A code both for acquisition and for coarse position information—there is no hand-over.

Time Measurements

Once the NAVSTAR/GPS receiver/computer has acquired the code (i.e., has synchronized itself as closely as possible with the satellite code), it can then measure the elapsed time since transmission by comparing the phase shift, or remaining difference, between the two codes: the more the two disagree, the greater the length of time since transmission, and length of time since transmission multiplied by the speed of light equals distance. For instance, if the receiver/computer determines that the closest the two codes can be synchronized is .07 seconds, then multiplying .07 by 162,000 nautical miles per second results in a distance of 11,340 nautical miles between satellite and receiver. Thus, by measuring phase shifts in the code, distance between satellite and receiver can be computed. Since the P code is 10 times more precise than the C/A code (10,230,000 bits versus 1,023,000 bits), it can measure the phase shift 10 times more precisely, which is why the P mode is so much more accurate than the C/A mode.

Position Fixing

If we know what our distance is from a specific point in space (a satellite in this case), then it follows that we must be located somewhere on the surface of a sphere with a radius of that distance, centered on that point. Figure 12-1 illustrates the aircraft position on the surface of a sphere (for clarity, only the relevant part of the sphere is shown). The first distance measurement then gives us our first Line of Position (actually a Surface of Position—an LOP does not have to actually be a line).

With two satellites and two distances, position can be further refined to the intersection of two spherical surfaces, which is a curved line. (Actually, the intersection of two spheres is a circle, but the NAVSTAR/GPS computer uses initial position information to reduce the circle to a curved line.) A third satellite provides a third surface, and the intersection of that surface with the previously established curve is a point in space based on the distance in nautical miles between the aircraft and each of the satellites (See Fig. 12-2).

Knowing that you are 13,000 nautical miles from satellite #1, 14,000 nautical miles from satellite #2, and 15,000 nautical miles from satellite #3 is not in itself useful information, especially since each of the satellites is traveling at 7,500 NMPH and is no longer where it was a second ago. This problem is solved by having each satellite transmit its position—its *ephemeris*—to the receiving computer, so that it can translate those distances into useful information: longitude, latitude, and altitude. These computations are enormously complicated and involve matrix algebra and the solution of simultaneous equations with four unknowns, but that is the sort of thing computers do well, for which we can be grateful.

Receiver Clock Correction

Since only the satellites are equipped with atomic clocks, there is an inevitable error in the measurement of the elapsed time between signal transmission and reception that is caused by the relative inaccuracy of the airborne quartz clock. This error is called the *time bias*, and because of it, range (distance) measurements are not as exact as we might like, and are called *pseudo range* estimates, as opposed to true range estimates. (The surfaces that result from pseudo range measurements are not true surfaces, but have, in fact, a certain "thickness.") If we were to use pseudo range measurements directly, we would end

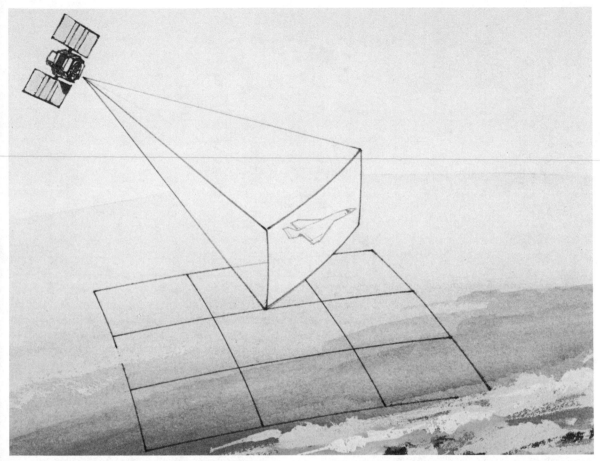

Fig. 12-1. Distance from one satellite provides aircraft position somewhere on the surface of a sphere. Courtesy of Rockwell International.

up not with an actual point fix, but with a three dimensional area fix, and accuracy would be seriously jeopardized. To correct for the time bias, a fourth satellite is used to provide a fourth LOP, which eliminates the time bias and reduces the area fix to a point in space. In practical terms, the fourth satellite substitutes for the missing atomic clock on the receiving end.

It is interesting to note that the basic satellite position computation can also be worked backwards to provide a precise time measurement. That is, if you provide the computer with your exact position, it can then solve for the exact time. This capability will be very useful to scientists and engineers in fixed locations who require highly accurate time checks. It may also be useful in air navigation at some point in the future, since accurate time checks can be used to improve the accuracy of other systems. (More on this in the section on integration of NAVSTAR/GPS with LORAN-C and INS.)

Ionospheric Propagation Error

The ionosphere refracts UHF satellite transmissions in the same way it refracts VLF, L/MF, and HF transmissions (only less so). Since a refracted signal has further to travel than a straight one, it will arrive later, distorting the time measurement, which will distort the distance measurement. If the ionosphere were always constant, this factor could be accounted for in the computer, but we know from its affect on NDB, LORAN-C, and OMEGA/VLF propagation that the ionosphere is hardly constant—

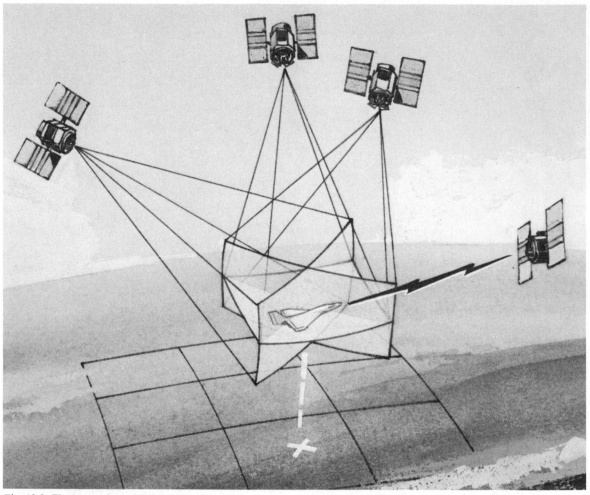

Fig. 12-2. The intersection of the surfaces of three spheres results in a fix in three dimensions. A fourth satellite is used to correct for clock errors. Courtesy of Rockwell International.

rather, it is constantly changing with the time of day, the season, and the degree of solar activity. Time measurements are too critical in the case of NAVSTAR/GPS positioning to ignore this ionospheric distortion, as is done with NDB, or to attempt to predict, as is done with LORAN-C and OMEGA/VLF; a better method of compensation is needed.

The ionosphere refracts, or delays, signals in an amount that is inversely proportional to the square of their frequencies, which means that the higher the frequency, the less the delay, at a rapidly decreasing rate. NAVSTAR/GPS satellites transmit on two different UHF frequencies (1575.42 MHz and 1227.60 MHz), as a result of which each is affected by the ionosphere to a different degree. By comparing the phase shift between the two frequencies the amount of ionospheric distortion can be measured directly. (For example, if there is no atmospheric distortion, both signals will arrive at the same time, but if there is a large amount of atmospheric distortion, the phase shift will be proportionally large.) The exact correction factor necessary for the conditions that actually exist at the time of satellite transmission can then be entered into the computer, and ionospheric propagation error can be effectively canceled.

It is easy to see from the preceding discussion

that not only did a satellite system of navigation have to wait for space technology to reach a certain level of sophistication, but it also had to wait for the development of the atomic clock and the compact, high speed, digital computer. In fact, the most limiting factor is the computer: a system at least as accurate as OMEGA/VLF could probably have been developed with earlier space technology and quartz clocks, but without compact, high speed digital computers, practical satellite air navigation could not have existed.

DERIVATIVE INFORMATION

NAVSTAR/GPS is much like LORAN-C, OMEGA/VLF, and INS in that it is primarily a position determining system, with all derivative information being obtained by observing changes in position. NAVSTAR/GPS determines actual track, for instance, from a series of position fixes. It determines ground speed by calculating the distance between two fixes and then measures the amount of time between the two to obtain a rate. It determines drift angle by comparing aircraft heading with actual track. (NAVSTAR/GPS requires heading and airspeed information to determine certain parameters—cross track error, drift angle, wind direction and velocity—just as LORAN-C, OMEGA/VLF, and INS do.) All of the derivative information common to all long range, area navigation systems will also be available with NAVSTAR/GPS, and the practical applications for NAVSTAR/GPS will be identical to those for LORAN-C, OMEGA/VLF, and INS.

NAVSTAR/GPS does have the capability to determine velocity directly (rather than by calculating the rate of change). It does this by measuring the Doppler shift in the satellite carrier frequency: Doppler shift is a direct function of relative velocity—satellite versus receiver—and since the satellite velocity is known, any further shift has to be a function of aircraft velocity.

At this stage of development, it is anticipated that most systems will not incorporate Doppler measurements, but will instead calculate rate measurements in the conventional manner. It is possible that the Doppler method may have advantages that are presently unknown, and this capability may take on more importance in the future.

PROTOTYPE USER SYSTEMS

A complete NAVSTAR/GPS navigation system requires an antenna, a Receiver/Processor Unit (RPU), and a Control-Display Unit (and, of course, the necessary connections). NAVSTAR/GPS RPUs can be made small enough and light enough to fit into panel-mounted units with built-in CDUs (although none has been developed to date). Remote units housed in ARINC boxes have been developed, but are not yet commercially available. The output of these remote units will either be directed to dedicated NAVSTAR/GPS CDUs or to the CDUs of existing long range systems, such as INS, that have the capability to accept NAVSTAR/GPS inputs. Most existing and all future navigation management systems will also accept NAVSTAR/GPS output. (See Chapter 17, Navigation Management Systems.) A complete family of military NAVSTAR/GPS systems (the only type currently in production) is illustrated in Fig. 12-3.

Antenna requirements are expected to be minimal: NAVSTAR/GPS antennas are easily mounted, their placement and installation is not especially critical, and they are low in drag and cost. (See Fig. 12-4.) Operational testing for military NAVSTAR/GPS systems is presently underway, with civilian testing to follow. Examples of installations on Navy and Air Force aircraft are shown in Figs. 12-5 and 12-6.

The flexibility inherent in NAVSTAR/GPS in terms of system weight and size, and the ease of installation of system components, promises to make the transition to NAVSTAR/GPS a relatively painless one.

INTEGRATION WITH LORAN-C AND INS

Because of the constantly changing relationship in satellite orbit patterns, there will be times each day when satellite geometry is adequate but not optimum. During these periods NAVSTAR/GPS accuracy will be somewhat degraded. By integrating NAVSTAR/GPS with other long range systems, such as LORAN-C and INS, overall system accuracy can be maintained at a high level during these periods of less than perfect satellite geometry.

Conversely, since NAVSTAR/GPS is capable of

Fig. 12-3. These Collins NAVSTAR/GPS systems will be manufactured by Rockwell International for use in a variety of military ground, airborne, and sea applications. Included is a manpack and vehicle adapters for jeeps and tanks (upper right), five Receiver/Processor Units (center row), a console mounted aircraft Control-Display Unit (center foreground), antenna assemblies (center foreground and rear) and ship's mounting units (upper left). Courtesy of Rockwell International.

determining velocity with considerable accuracy—better than one nautical mile per hour—rate aiding of LORAN-C with NAVSTAR/GPS rate information will greatly improve the accuracy of this system. (NAVSTAR/GPS can be used to provide rate aiding information for OMEGA/VLF as well, however OMEGA/VLF has little to offer NAVSTAR/GPS in return. The two systems are not mutually beneficial—OMEGA/VLF-NAVSTAR/GPS integration is a one-way street.)

NAVSTAR/GPS will be able to provide INS systems with continuous updates, bounding drift to less than 0.1 nautical miles. The military is taking this one step further, researching ways of integrating NAVSTAR/GPS with INS to produce fixes that are far more accurate than the best either is capable of independently.

During times of high-vehicle dynamics, for instance (which means tactical maneuvers at high speeds and high G loadings), NAVSTAR/GPS accuracy begins to breakdown; during these times of high-G loading, INS can be used to provide the NAVSTAR/GPS computer with the loading information necessary to maintain the desired accuracy. So far, total NAVSTAR/GPS-INS integration is of concern only to the military, but it points towards areas where military needs and research may ultimately provide benefits for civilian aviation also.

Finally, since NAVSTAR/GPS is also capable of providing extremely accurate time checks (given

Fig. 12-4. A 14-inch diameter white NAVSTAR/GPS antenna is shown here on the top of a US Navy A-6E aircraft, between the canopy and the tail assembly. Courtesy of Rockwell International.

a known position), it is possible to take the NAV-STAR/GPS time signal and use it to synchronize all LORAN-C transmitters and receivers, which would allow LORAN-C to function in a direct measurement mode, just as NAVSTAR/GPS does. This would greatly increase the accuracy of the LORAN-C system. Thus, during periods of poor satellite geometry, the time standard provided by the satellite would enable the LORAN-C system to function well enough to maintain a constant level of accuracy.

Many experts in air navigation predict that an integrated LORAN-C-NAVSTAR/GPS navigational system may prove to be the best overall system for civilian aviation, while the optimum military system will probably be an integrated INS-NAVSTAR/GPS system. If this proves to be the case, it could well force the extension of the LORAN-C system beyond its present limits.

ADDITIONAL CAPABILITIES

If NAVSTAR/GPS only provided universal, random route navigational guidance to within 100 meters, that would be justification enough for the system, but in fact, it has capabilities even beyond that. One such possibility is *Differential GPS*.

Differential GPS is a system addition in which a ground relay station continuously receives and corrects the satellite signal, and these corrections are then passed on to all users in the area. The result is even greater accuracy than is possible with direct access to the satellite information. Situations where this capability might be useful would be in precision approach situations, where a higher degree of accuracy might even allow zero-zero approaches (CAT IIIC approaches), and in terminal control operations, where increased accuracy would allow tighter control with resulting improvements in traffic flow and separation. Even without Differential GPS aiding, NAVSTAR/GPS will almost certainly be able to provide accurate approach information to at least non-precision standards for every IFR airport in the world, and three-dimensional NAVSTAR/GPS approaches are within the realm of possibility.

Another area for possible expansion is in data transfer. A portion of the navigation message transmitted by each satellite (which includes system status, ephemeris, clock corrections, and ionospheric corrections) has been reserved for data transfer. This could be used by ATC for clearances, for the transfer of important weather information, for other information of a time critical nature, or for any number of other uses—there will undoubtedly be a period of experimentation before the most appropriate use of the data transfer portion of the navigation message has been determined, but the capability is built into the system.

A further expansion would be the transfer of data from the user—the aircraft—back to the satellite for relay to a ground station. This would change the system from a passive to an active system, but would allow, for instance, for the automatic reporting of altitude and position back to the air traffic control system. (The military would no doubt want to retain the option to switch between passive and active as the tactical situation dictated.) Automatic reporting of altitude and position would greatly improve ATC's ability to control and separate aircraft, and terminal procedures, in particular, would be greatly expedited.

Fig. 12-5. A technician works under a NAVSTAR/GPS equipped A-6E onboard the aircraft carrier USS America. The Collins NAVSTAR/GPS receiver is located in the radio rack accessed through the open hatch under the fuselage. Courtesy of Rockwell International.

TIME TABLE FOR COMPLETION

Ten fully operational NAVSTAR/GPS satellites are presently orbiting the earth, providing three-dimensional coverage for about five hours each day in an area near Yuma, Arizona. This is sufficient for testing and evaluation, but is obviously unsatisfactory for navigation. The remainder of the satellites—eight plus three spares for a total of 21—were sched-

uled to be launched from the space shuttle at the rate of seven per year beginning in 1987, with worldwide two-dimensional coverage expected in the first part of 1988, and worldwide three-dimensional coverage planned for late 1988. With the space shuttle now grounded (at the time of this writing), and with two consecutive failures of the unmanned missiles that were intended as substitutes, it seems very unlikely

Fig. 12-6. The first B-52G aircraft to be equipped with a NAVSTAR/GPS receiver is shown following installation at Tinker Air Force Base, Oklahoma. The 14-inch diameter NAVSTAR/GPS antenna is located approximately 42 feet from the nose of the aircraft, as indicated by the arrow. Courtesy of Rockwell International.

that this schedule can be maintained. I think it is safe to say that the NAVSTAR/GPS system will not be delayed indefinitely or scrubbed altogether, but I think it is also safe to say that no one knows for sure when it will be complete at this time. (Perhaps the schedule will be better known by the time you read this.) However long it takes, it should be worth the wait.

THE DIMINISHING ROLE OF THE VOR

What does NAVSTAR/GPS mean for the VOR/DME network? My guess is that VOR/DME navigation actually began to be obsolete with the rapid acceptance in the past several years of long-range area navigation at all levels of aviation, from LORAN-C for private general aviation through OMEGA/VLF and INS for corporate and airline aviation. In practical terms however, NAVSTAR/GPS is probably the only system capable of actually replacing VOR/DME as the primary navaid in the National Airspace System. The VOR/DME system, with its hundreds of ground based facilities, each sitting on its own site and requiring constant monitoring and maintenance, is very expensive, and the favorable cost advantage alone of a NAVSTAR/GPS based NAS versus a VOR/DME based NAS should make that transition inevitable.

The inevitability of a satellite system for primary navigation does not necessarily mean that the transition will be a rapid one though: the change from an airway to an area-based system of aircraft control and separation is a much more complex task than simply requiring all IFR approved aircraft to have NAVSTAR/GPS receivers, as we have seen throughout these chapters covering area navigational systems. New policies have to be formulated, computers must be upgraded and completely reprogrammed, controllers must be retrained, standard operating procedures and letters of agreement between sectors must be redeveloped, and regulations and guidelines must be rewritten before this can happen. Nonetheless, barring further significant setbacks in the NAVSTAR/GPS program, the change from a VOR/DME-based airway system to a satellite based random route system would seem to be an inevitability.

THE LONG-TERM OUTLOOK
FOR GENERAL AVIATION

The long-term outlook for general aviation could not be better as regards to air navigation. NAVSTAR/GPS promises to provide low cost, highly accurate, universal, random route navigational guidance with very few limitations. NAVSTAR/GPS should be close to a perfect navigational system—it would be terrific if the P code could also be made available to civilian users, which would provide for precision approach guidance to at least CAT II standards, but NAVSTAR/GPS should still be an incredible system. If problems do arise, they are much more likely to be political or bureaucratic in nature than they are physical; none threaten at the present time.

Perhaps the best news for general aviation is that NAVSTAR/GPS will not be an exotic system costing hundreds of thousands, or even tens of thousands, of dollars. While manufacturers are reluctant to commit themselves to precise cost estimates at this stage, it would appear that NAVSTAR/GPS systems can be built and installed for somewhere around $4,000.00 for entry-level systems. While this estimate may be a little optimistic, even if it is in error by half, clearly this is not an exotic system in terms of cost.

Volume production would reduce costs further, and volume production is, in fact, a very real possibility: the automobile industry is investigating the practicality of installing satellite receivers in automobiles, and the shipping, rail, taxi and security industries are also very interested in NAVSTAR/GPS. If NAVSTAR/GPS were to become a household word, with a satellite system offered as an option in virtually all new cars plus additional units sold to a variety of other users, the production levels necessary to satisfy this demand would inevitably lead to a reduction in cost for airborne units as well.

There is, in short, every reason to believe that NAVSTAR/GPS will be the primary system of navigation for all aircraft within just a few years, and that it will continue to be the primary nav system for at least the rest of this century, and possibly for well into the future.

Chapter 13

Over-Water Navigation

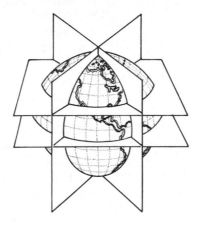

Most air navigation and communication is short range and line-of-sight, utilizing VHF and UHF frequencies. This has shaped the way most navigation and air traffic control is conducted: domestic air navigation tends to be VOR/DME airway navigation, with radar supervision and control for back-up and expediency wherever possible. These kinds of facilities and this kind of navigation and control is not possible over water. As a result, over-water navigation tends to be something that is quite different from land navigation.

Over-water navigation will always be a little different from land navigation in any case—the ocean offers few alternatives when fuel runs short, or navigation proves to be inexact, or mechanical problems arise. Even if navigation were the same over water as it is over land (as it may well be in the future when NAVSTAR/GPS is fully operational and has been completely integrated into the NAS), over-water operations would still be different from land operations, for a variety of reasons.

Many of the characteristics and techniques of over-water navigation have been covered in this book, indirectly, in the chapters covering long-range navigation systems. The specifics of over-water navigation are worth a separate look however, and that is the subject of this chapter.

This chapter will concentrate on the navigational aspects of over-water operations, but communications, air traffic control, customs, immigration, agriculture, health, survival preparations, and aircraft contingency planning are inevitably involved and intertwined with the navigational aspects of international travel. These other elements will be touched on in this chapter from time to time, but the emphasis in every case will be on navigation.

This chapter is not meant to be a complete preparation for over-water flight—that would take a good-sized book in and of itself. Over-water and international requirements change on a daily basis in any case, and any attempt to be complete would inevitably contain many outdated elements by the time you were to read it. The purpose of this chapter is to provide an overview of international flight planning and an introduction to over-water navigation. Additional sources of information for those contemplating actual over-water flights are provided in Appendix A, Additional Sources Of Information, and Appendix B, Charts and Chart Sources.

GENERAL CHARACTERISTICS

Over-water navigation is inherently different from airway, and land navigation. Even though many of these differences are fairly obvious, they are still worth examining briefly in order to understand the specific problems involved.

The first and perhaps most obvious difference between over-water navigation and land navigation is that over-water navigation is generally done without the benefit of any contact with the terrain: pilotage is impossible over water, and over-water navigation is, for all practical purposes, the equivalent of continuous instrument navigation. Some over-water routes do, of course, provide some terrain contact, such as many of the Caribbean routes and the most northerly of the North Atlantic routes, and some routes do provide for an occasional ground check (weather permitting)—Gander to Iceland via the southern tip of Greenland, for instance—but over-water navigation is generally done without the benefit of terrain reference.

In theory, this shouldn't matter—modern air navigation has very little to do with pilotage—but we all know that in practice it does. Human beings like ground contact, and they tend to rely on it when they have it, and they tend to be less secure in their estimates of position, time, and track when they do not.

Second, over-water navigation generally provides for few if any emergency landing alternates. This not only makes fuel planning much more critical, but it raises the risks associated with in-flight emergencies and abnormalities. Many emergencies and abnormalities can be best solved, or perhaps only solved, by landing as soon as possible—engine fires, hydraulic leaks, fuel leaks, structural failures, severe icing, and so on. This option is not available overwater. (Ditching is always an option of course, but not a good one.)

Third, almost all over-water navigation is done out of radar contact. (See also the next chapter, Navigation In Non-Radar Environments.) Over-water navigation is *strategic navigation*; that is, it relies on fundamental principles of dead reckoning to maintain aircraft separation, and aircraft separation over the high seas is a cooperative venture—safety is a direct function of the navigational accuracy and care taken by all involved. Even position reporting is inherently different out of range of land based aids, since it is based on estimates of position based on long range navaids, rather than positive passage over ground based facilities. Reporting your position as "51 North, 30 West", for instance, based on a long-range system such as OMEGA/VLF or INS, is not the same as reporting over the GANDER VOR—a known fix on the ground with an unmistakable station passage. These differences are taken into account by those responsible for the navigation and separation standards along over-water routes, and greater tolerances are accordingly built into the system, but it is important that pilots understand the differences and their significance.

Another difference between land and water navigation is that the pilot must rely on long-range systems over water, whereas he normally uses long-range systems only as a supplement and back-up over land. The only back-up to a long range system over water (other than another long-range system) is dead reckoning, an inherently less effective back-up than any electronic navaid. Over-water navigation is therefore, for most pilots, a new and somewhat strange affair.

Finally, most over-water navigation either begins or ends in a foreign country. Foreign countries have different languages, air traffic systems, regulations, facilities, and entry and exit requirements. This increases both the workload and the stress—accents interfere with communication, navigational procedures can be very different (the metric system is used internationally for nearly all aspects of flight except airspeed and altitude)—and regulations, restrictions, and bureaucrats abound. This is not to say that international travel is impractical, dangerous, or not worth the effort, but the demands are greater.

Some of the specific ways of dealing with the differences just discussed are covered in the following sections.

NAVIGATIONAL REQUIREMENTS

Many over-water routes do not have specific electronic navigational requirements. Much of the low-altitude, over-water airspace is uncontrolled, both VFR and IFR, and pilots willing to fly low enough to avoid controlled airspace can, in theory, navigate over large bodies of water using only dead reckoning and those VOR or NDB aids as happen to be available.

In controlled airspace, ICAO standards require only that the aircraft be able to navigate with sufficient accuracy to be able to comply with the clearance. (This is similar to the FAA regulation that requires that IFR communication and navigation equipment be appropriate to the ground facilities to be used. [FAR 91.33 (d) (2)].) Unless specifically stated otherwise, the equipment required depends on the type of route and the separation standards applied to that route.

While it is still theoretically possible to navigate over much of the globe using only dead reckoning and NDB, for all practical purposes the absolute minimum in long-range equipment for most over-water operations is a single LORAN-C system. Most pilots who regularly fly over water want two long-range systems, at least one of which is a worldwide system such as OMEGA/VLF or INS, and for dispatch reliability a third system is a necessity. (High-Frequency radio is also both a practical and a legal necessity; this is covered later, in Communications.)

Over the North Atlantic, specific navigational requirements do exist, both for turbine aircraft at the normal cruising Flight Levels, and for all aircraft departing from or flying through Canadian airspace. For filing along the North Atlantic Organized Track System (NAT OTS), the aircraft must be equipped with dual long-range systems and meet and be certified for ICAO Minimum Navigational Performance Specifications (MNPS). Without proper certification, the flight will not be allowed to enter the MNPS airspace, which includes most of the North Atlantic between FL 275 and FL 400. Special North Atlantic routes off the Organized Track System, both high and low altitude, are available for aircraft not meeting the full MNPS requirements, however these routes are less direct. The western halves of some of these special routes are illustrated in Fig. 13-1.

Pilots of single-engine aircraft en route to Europe from or over Canada must land at Moncton, New Brunswick and be checked for a long list of survival and basic navigational gear, as well as for knowledge of fundamental long-range navigation and communications. (Your local GADO or FSS can put you on the right track to prepare for this inspection.) This check must be done regardless of the route or altitude proposed.

In addition to the obvious need for adequate long-range electronic guidance over water, basic instrumentation must also be complete and in top working condition for any over-water flight. Compasses should be swung and checked for leaks and cracks, clocks should be calibrated and set to Coordinated Universal Time (UTC), gyros should be in top working order, vacuum systems should be inspected, and pitot-static systems and altimeters should be tested. Batteries and generators should also be inspected and charged or replaced as necessary. All of these components are essential for dead reckoning, and dead reckoning is an essential part of over-water navigation—no sensible pilot relies exclusively on his long-range nav system or systems.

FLIGHT PLANNING

Flight planning for over-water operations is quite a bit more complicated and more critical than it is for land operations: fuel planning in particular is very important, and the ICAO Flight Plan is different from and more complicated than the FAA Flight Plan.

Preliminary Planning

Flight planning for an over-water flight to a foreign destination normally must begin weeks, even months, in advance. The first step in the preliminary planning is to consult the *International Flight Information Manual* (IFIM), published by the FAA in cooperation with ICAO and participating governments, to determine passport, visa and immunization requirements. Entry and exit permits are often required, as are separate landing permits; refueling restrictions also exist at many destinations. (The IFIM is available through the U.S. Government Printing Office, Appendix A.)

Fig. 13-1. Pilots operating below MNPS airspace have a variety of routes between the Canadian coast and Iceland available to them. Similar routes exist between Iceland and Europe. Reproduced with permission of Jeppesen Sanderson, Inc. Not actual size. Not to be used for navigation.

As a part of the preliminary planning, fuel and navigational requirements for the preferred route should be considered. The best routes are usually the ones with the longest range requirements, for which auxiliary fuel tanks may have to be installed. Auxiliary fuel tanks not only take time to install, but also require thorough testing and certification by an aircraft mechanic with an Inspection Authorization. Navigational requirements will be directly related to the route selected and the degree of navigational redundancy desired: in most cases at least a single LORAN-C will be required, and this equipment should be ordered and installed well in advance.

Survival gear must also be obtained, and arrangements must be made at both ends for weather briefings, for the filing of flight plans, for fuel purchases (many foreign destinations do not have FBOs as we know them), and even for parking. Very often these arrangements involve contracting with a commercial agent at the destination who handles these details for the pilot—in some cases an agent is even required.

The Aircraft Owners and Pilots Association (AOPA) can provide assistance in these areas, as can commercial organizations specializing in international operations such as Universal Weather and Aviation, and Lockheed DataPlan (Appendix A). The pilot planning on making a crossing for the first time would be well advised to take advantage of one of these services—there is almost no other way the inexperienced or occasional over-water operator can ensure that all current requirements have been satisfied.

Charts

Both Jeppesen and the National Ocean Service (Appendix B) can supply all the necessary charts for over-water navigation. In addition to the instrument route and orientation charts, a variety of other charts are available for terrain orientation and emergency use. World Aeronautical (WAC) and Operational Navigational Charts (ONC) are the most common, but Jet Navigation Charts (JNC), Global Navigational Charts (GNC), and LORAN-C Navigational Charts (LCC) are also available (Appendix B). A section of a Jeppesen long-range instrument chart, AT (H/L) 1, showing common Canadian arrival and departure

fixes directly to and from Europe, is illustrated in Fig. 13-2.

Flight Logs

Standard IFR flight logs do not work very well over water: standard logs are predicated on airway and navaid fixes (rather than on lat/long coordinates), and they very often do not include columns for supplemental dead reckoning information. (A dead reckoning log should be maintained over water as a back-up and for verification of the proper functioning of the primary long-range systems.)

Figure 13-3 shows a military version of an over-water flight log which can be adapted quite easily to civilian use. Most of the columns are fairly self-explanatory; note that the route block is extra large to allow for full lat/long coordinate entry (L/O stands for Level Off), and note also that a significant portion of the flight log is dedicated to dead reckoning computations. A student pilot would have no trouble completing this log, beginning with T.C. or G.C. (True Course or Grid Course—the military uses Grid Navigation in polar regions; this can be disregarded for most civilian operations), and continuing on through columns for winds aloft and drift correction, true heading, variation, magnetic heading, magnetic course, temperature and altitude, TAS, ground speed, distance, estimated leg times, and accumulated time.

Two columns are provided on this log for ETAs—an excellent idea. The first column is for initial, preflight ETAs, and the second column is for revised ETAs en route, using computed ground speed instead of estimated ground speed. A block for ATA over each fix is included as well, directly under the revised ETA (RETA). The last column before the remarks column can be used for estimated and actual fuel remaining over each fix, or fuel can be monitored separately with a range control graph (discussed later in the chapter).

FUEL PLANNING

The section on the far right of this flight log form is a summary of fuel planning information. For short-range over-water flights where the computation of en route fuel is fairly straightforward, fuel planning

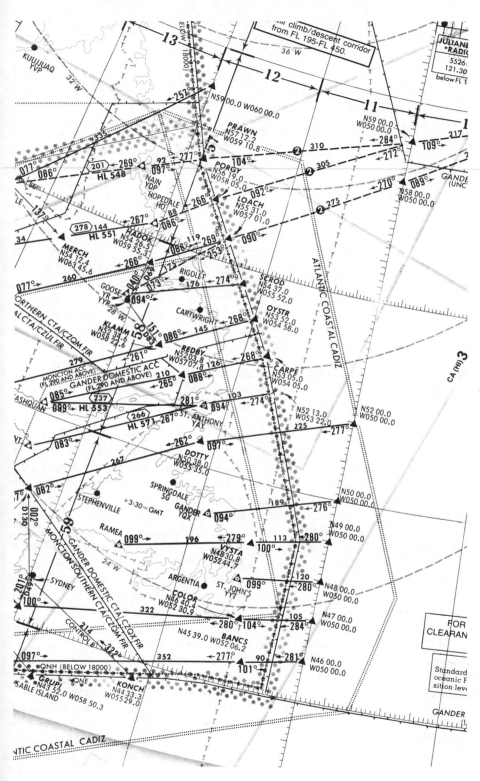

Fig. 13-2. Arrival and departure fixes for the North Atlantic Organized Track System surround the Canadian coast. Reproduced with permission of Jeppesen Sanderson, Inc. Not actual size. Not to be used for navigation.

MISSION PREFLIGHT LOG	NAVIGATOR: Lt Luhrman	TRAINING MISSION COMMANDER: Capt McNeely	AIRCRAFT COMMANDER: Maj Neveng	MISSION: 51-40	OPERATING WT: 66,400	CLIMB TEMP DEV: +5	ACFT CALL SIGN: Gator 79
DESTINATION: Wake Is AFB N19-17.0 E166-380	ROUTE: MDWAKATC	SQUADRON: 452FTS	CLASS:	CREW: N01	SCHEDULE/ACTUAL TAKEOFF: 1900 Z/	Z DATE OF TAKEOFF: (Z) 2Jul79	CARGO/PAX WT: 4,600	CRUISE TEMP DEV: +1	ACFT SERIAL NO: 1404

Ramp fuel wt: 37,000 Takeoff wt: 107,000 Acft type: T43A
Ramp gross wt: 108,000 PG: 4-6

ROUTE IDENT	FROM Midway NS / ROUTE	T.C. OR G.C.	W/V DC	T.H. OR G.M. VAR OR GRIV	M.H.	(M.C.)	TEMP ALT / TAS TRUE MACH	GS	GRD DIS ACC. GRD DIS	TIME ACC. TIME	PROP ETA	RETA ATA	A OR B	REMARKS
A	NQH 220/112 N27-00.0 W179-07.0	230	300/10 +1	231	-10	221	220 / M 360	357	112 112	+18 +18				
	L/o	273	300/10 +1	274	-10	264	263 / M 360	357	6 118	+01 +19				AD116
B	N27-02.0 W180-00.0	273	320/10 +1	274	-10	264	263 -45 426 31.0M 172	420	47 165	+07 +26				Dept
C	N27-03.0 E175-00.0	271	320/10 +1	272	-9	263	262 ✓ M	420	267 432	+35 1+05				
D	N27-00.0 E172-00.0	270	320/10 0	270	-7	263	263 ✓ M ✓	421	160 592	+23 1+28				TP
E	N26-00.0 E169-55.0	224	330/10 0	224	-6	218	218 ✓ M ✓	428	164 756	+23 1+51				
F	N21-00.0 E166-00.0	223	360/05 0	223	-6	217	217 ✓ M ✓	430	323 1079	+45 2+36				Term
G	AWK N19-18.0 E166-36.0	162	010/05 -1	161	-6	155	156 ✓ M ✓	430	107 1186	+15 2+51				
			/				M							
			/				M							
			/				M							
			/				M							
			/				M							
			/				M							
			/				M							
			/				M							
			/				M							
			/				M							
	TERMINAL ALTERNATE: NR		/				M						A.D. =	

Fuel planning summary:

	TIME	FUEL
1. ENROUTE	2+51	16.7
2. RESERVE	+20	4.0
3. ENROUTE PLUS RESERVE	3+11	20.7
4. ALTERNATE (AND MISSED APPROACH)	NR	NR
5. HOLDING	—	—
6. APPROACH/LANDING	+15	1.0
7. IDENTIFIED EXTRA	—	—
8. TOTAL (3+4+5+6+7) TAKEOFF/FLAPS UP	3+26	21.7
9. TAXI AND RUN UP (Acceleration)		1.0
10. REQUIRED RAMP		22.7
11. ACTUAL RAMP		
12. UNIDENTIFIED EXTRA		
13. REQUIRED OVER DEST (4+5+6)		1.0
ENDURANCE HRS	BURN OFF (1+6+7) 17.7 LBS	
HIGHEST ACCEPTABLE FL	LOWEST ACCEPT. FL	
WIND FACTOR: -2	TOTAL -5	1ST HALF END HALF +3
TOTAL DISTANCE ( )	1 T ( )MIN	
(WF2-WF1)÷2 (TAS)( )		
TOTAL TIME - T = TIME TO ETP		

Fig. 13-3. A flight log of the type commonly used for over-water navigation. This log is for a flight between the Pacific islands of Wake and Midway.

can be accomplished directly on the flight log using this column. For long-range over-water flights, where range is a critical factor, particularly for those flights involving substantial changes in the forecast winds aloft over the course of the flight, more detailed en route fuel planning is necessary.

Figure 13-4 illustrates a fuel planning worksheet appropriate for long-range, over-water flights. The right-hand column of this worksheet is essentially the same as the fuel planning summary on the flight log. The bottom half of the form provides room for position reports and clearances, and the top left-hand side is for detailed fuel planning by en route segments.

This particular worksheet has room for five en route segments: climb plus four cruise segments. (A descent segment could also be included; however, conservative fuel planning treats the descent as part of the last en route segment, and any gains in fuel consumption that occur during the descent are used to increase the reserve fuel at the destination.) Four cruise segments allow for four different winds aloft

segments, or four different altitude and fuel flow segments, or any combination of each.

To complete the fuel planning worksheet, the pilot divides the trip into a climb segment, plus as many cruise segments as are necessary due to significant changes in ground speed or fuel flow—four will normally be enough, and often as few as two will suffice. Estimated en route times are calculated for each segment and entered in the first column. These times are added to each other to get accumulated time (Accum Time). Total time is transferred to the top of the Time Remaining column. Each segment time is then subtracted from the time above it to produce time remaining at the end of each segment.

Once segment times have been computed, fuel consumption can be calculated for each segment. These cumulative amounts are then entered in column B, Fuel Consumed, and Fuel Remaining (column C) is entered by subtracting fuel consumed at the end of each segment from total fuel available at takeoff (subtract column B from column A). Total en route

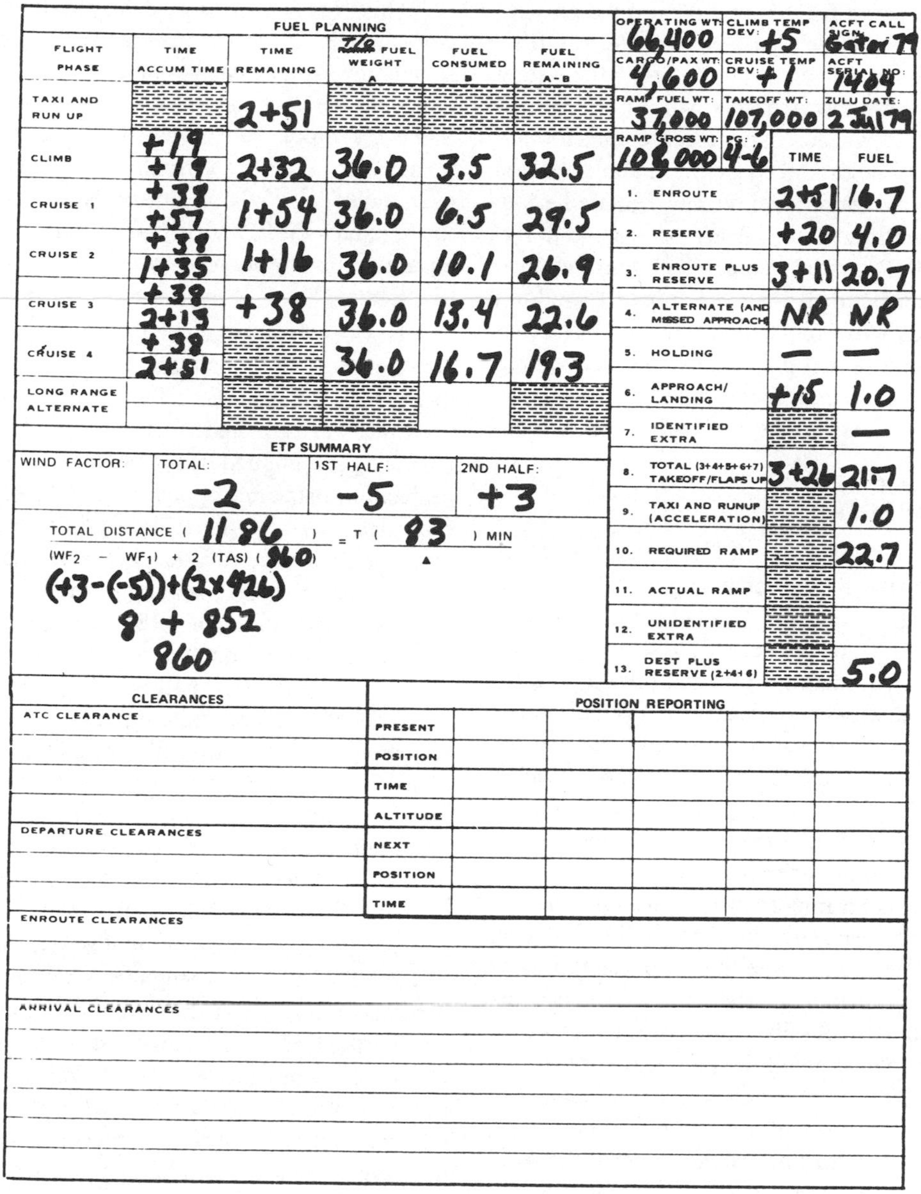

FUEL PLANNING

FLIGHT PHASE	TIME ACCUM TIME	TIME REMAINING	FUEL WEIGHT A	FUEL CONSUMED B	FUEL REMAINING A-B
TAXI AND RUN UP		2+51			
CLIMB	+19 / +19	2+32	36.0	3.5	32.5
CRUISE 1	+38 / +57	1+54	36.0	6.5	29.5
CRUISE 2	+38 / 1+35	1+16	36.0	10.1	26.9
CRUISE 3	+38 / 2+13	+38	36.0	13.4	22.6
CRUISE 4	+38 / 2+51		36.0	16.7	19.3
LONG RANGE ALTERNATE					

OPERATING WT: 66,400	CLIMB TEMP DEV: +5	ACFT CALL SIGN: ... 79
CARGO/PAX WT: 4,600	CRUISE TEMP DEV: +1	ACFT SERIAL NO: 1484
RAMP FUEL WT: 37,000	TAKEOFF WT: 107,000	ZULU DATE: 2 Jul 79
RAMP GROSS WT: 108,000	PG: 4-6	

	TIME	FUEL
1. ENROUTE	2+51	16.7
2. RESERVE	+20	4.0
3. ENROUTE PLUS RESERVE	3+11	20.7
4. ALTERNATE (AND MISSED APPROACH)	NR	NR
5. HOLDING	—	—
6. APPROACH/LANDING	+15	1.0
7. IDENTIFIED EXTRA		—
8. TOTAL (3+4+5+6+7) TAKEOFF/FLAPS UP	3+26	21.7
9. TAXI AND RUNUP (ACCELERATION)		1.0
10. REQUIRED RAMP		22.7
11. ACTUAL RAMP		
12. UNIDENTIFIED EXTRA		
13. DEST PLUS RESERVE (2+4+6)		5.0

ETP SUMMARY

WIND FACTOR:	TOTAL: -2	1ST HALF: -5	2ND HALF: +3

TOTAL DISTANCE (1186) = T (93) MIN

(WF$_2$ − WF$_1$) + 2 (TAS) (860)

$$(+3-(-5))+(2\times426)$$
$$\frac{8 + 852}{860}$$

CLEARANCES

ATC CLEARANCE

DEPARTURE CLEARANCES

ENROUTE CLEARANCES

ARRIVAL CLEARANCES

POSITION REPORTING

PRESENT	
POSITION	
TIME	
ALTITUDE	
NEXT	
POSITION	
TIME	

Fig. 13-4. Fuel planning worksheets can be very helpful when planning long over-water flights.

fuel required is the last fuel consumed figure in column B, and this is the figure that is entered on Line 1 of the fuel planning section to the right. Since the total en route fuel required figure takes into account the climb segment plus the variation in cruise segments as ground speeds and fuel flows change, this will be a much more accurate figure than that obtained in the conventional manner by estimating an average ground speed and fuel flow for the entire flight. For long-range, over-water flights, this extra attention to detail is important.

Once total en route fuel required has been calculated, completing the rest of the fuel planning worksheet is fairly simple and straightforward. Item 2, for instance, is reserve fuel, Item 3 is en route plus reserve, Item 4 is missed approach and alternate fuel, and so on down the form. (Identified Extra—Item 7—means fuel for known delays, while Unidentified Extra—Item 12—means extra reserve fuel for unknown contingencies. All fuel carried above what is required—actual ramp fuel versus required ramp fuel—is unidentified extra.) The most important figure is Item 10, Required Ramp, which must be equal to or less than Item 11, Actual Ramp. (It wouldn't do to have less actual fuel than is required.) The conscientious use of a fuel planning worksheet will ensure that all fuel requirements have been met, and that actual fuel carried is adequate to cover all contingencies.

EQUAL TIME POINT

The *Equal Time Point* (ETP) is that point during the flight where time remaining to the first suitable airport after coast-in—landfall—equals time to return to the last airport before coast-out. Military controllers will sometimes give the pilot a time-to-dry-feet on initial radar contact, which is the amount of time it will take at that point to reach land in the shortest direction. The ETP is simply that point where the time-to-dry-feet is exactly the same no matter which way you turn: ETP translates into equal-time-to-dry-feet (or longest-time-to-dry-feet).

The ETP is an important over-water flight parameter: in an emergency or abnormality requiring a landing as soon as possible, it is important to know with certainty which direction to turn to reach an airport in the shortest possible time. At most points in the flight this decision will be obvious, but halfway between landfalls you could very easily make the wrong decision: time-to-dry-feet is a function of ground speed, and turning around reverses the wind component. The stronger the wind component, the more the ETP will vary from the midpoint.

There is, fortunately, a simple formula to determine the ETP, although it does require some explanation. The formula is:

$$\frac{\text{TOTAL DISTANCE}}{(\text{WF}_2 - \text{WF}_1) + 2 * \text{TAS}} = \frac{\text{ETP}}{60}$$

This formula is printed on the Fuel Planning worksheet in the block marked: ETP Summary. As shown, the formula can be solved directly on an E6-B computer. (Note that the formula printed in the ETP Summary block substitutes the E6-B hourly rate pointer for 60 in the formula above.) The formula even includes provision for two different wind factor segments (WF_1 and WF_2), although it will work equally well with a constant wind over the entire route of flight.

To work the formula on the E6-B, the difference between the second half and first half wind factors (headwinds are negative, tailwinds are positive), plus two times the estimated cruise TAS, is located on the inner ring and aligned with total distance between the nearest airports after opposite landfalls on the outer ring. The answer—the ETP—is then found on the outer ring over the pointer. In the example worked on the illustrated fuel planning form, with a total distance between airports of 1,186 nautical miles, a TAS of 426, and wind components of minus 5 knots (a headwind) for the first half and plus 3 knots (a tailwind) for the second half, the ETP turns out to be 83 minutes. This means that when 83 minutes remain in the flight to the first airport after normal landfall, then 83 minutes also remain to return to the first airport after the landfall behind. Prior to the ETP, the pilot should turn around to reach land soonest, and after the ETP, he should continue on.

The formula is simple, but it is easy to make mistakes with it: the wind components must be expressed in terms of negatives (headwinds) and positives (tailwinds), the first wind factor must be subtracted from the second wind factor ($\text{WF}_2 - \text{WF}_1$), and the algebraic rules of subtracting negative and positive numbers must be carefully observed (subtracting a negative is the same as adding a positive). Note that if the winds are assumed to be constant for the entire crossing, that the wind factors then cancel, which makes the formula very simple: a constant 50-knot headwind, for instance, becomes −50 minus −50, or zero. Finally, remember that the result, ETP, represents time remaining to reach the nearest airport after landfall in either direction—it does not mean elapsed time since the last landfall airport.

POINT-OF-NO-RETURN

The Equal Time Point is not the same as the Point-Of-No-Return (PNR). The ETP is that point where as much time remains to continue on as to turn around and return. The PNR is that point where you must continue on, because insufficient fuel remains to return. This is also an important over-water flight parameter, particularly when flying to island airports lacking alternates, or to any destination reporting deteriorating weather over a broad area. In these situations, the option to turn back is an important one; the PNR marks the point where that option is no longer available.

The PNR has to be calculated manually, by what is essentially a bracketing process. Once fuel remaining estimates have been calculated for various fixes en route, those estimates can be used to figure fuel remaining after return to the departure point, remembering to reverse the wind component after the turnaround. Once two adjacent fixes have been identified that bracket adequate and inadequate fuel reserves remaining after a return to the departure point, a more precise PNR can be determined either by interpolation or by further refinement between the two.

Exact precision in fixing the PNR is not so important as is having some idea where the PNR falls—if returning to the departure point becomes a real possibility en route, a precise PNR can be determined en route using actual fuel remaining and actual tailwind components. At a certain point fuel remaining will approach the minimum necessary to return: that is the last chance to make a decision to return.

SINGLE-ENGINE RANGE

An additional fuel planning consideration for multiengine aircraft is single-engine range (or one engine inoperative range for three- and four-engine aircraft). In most cases—there are a few exceptions involving four-engine aircraft—single engine range will be somewhat less than normal multiengine range. Fuel reserves must therefore be planned to allow for an engine failure at the most critical point, which is the ETP.

Once the ETP has been calculated, checking the single-engine range is very simple. The aircraft performance manual will list airspeeds and fuel flows for single engine operations, and, using that information (along with winds aloft for the single-engine ceiling, distance remaining at the ETP, and fuel remaining at the ETP), it is a fairly simple matter to ensure that fuel reserves are sufficient for continued single-engine flight from over the ETP. If conservatively calculated fuel reserves exist for multiengine flight, adequate fuel reserves will normally also exist for single-engine flight, since single-engine range is usually only slightly less than multiengine range. (The inefficiency caused by losing an engine is balanced by the increased efficiency that comes with slower, engine-out cruise speeds.) Single-engine range should not be taken for granted however, but should be verified prior to any over-water flight.

UNPRESSURIZED RANGE

A final fuel planning consideration, applicable only to pressurized aircraft, is unpressurized range. In many cases of cabin depressurization, descent to lower, survivable altitudes is necessary, even if supplemental oxygen is carried—many passenger oxygen masks are designed for emergency use only, and will not supply oxygen under enough pressure to ensure long-term consciousness. Even if the masks are capable of supplying oxygen at sufficient pressure to survive at cruise flight levels, it is still often not possible to carry enough oxygen to last for the duration of the flight if depressurization occurs at the ETP. If at all possible, sufficient fuel should be carried to descend to 12,000 feet at the ETP and continue at long-range cruise to landfall regardless of the oxygen supply. If this is not practical, then the pilot should ensure that fuel reserves are adequate to continue from the ETP at whatever altitude it is necessary to descend to after cabin depressurization, and that oxygen supplies are adequate for the length of time remaining for the altitude selected. It would be very unnerving to discover that the only alternative to unconsciousness was fuel exhaustion.

ICAO FLIGHT PLAN

The standard VFR/IFR flight plan that most U.S.

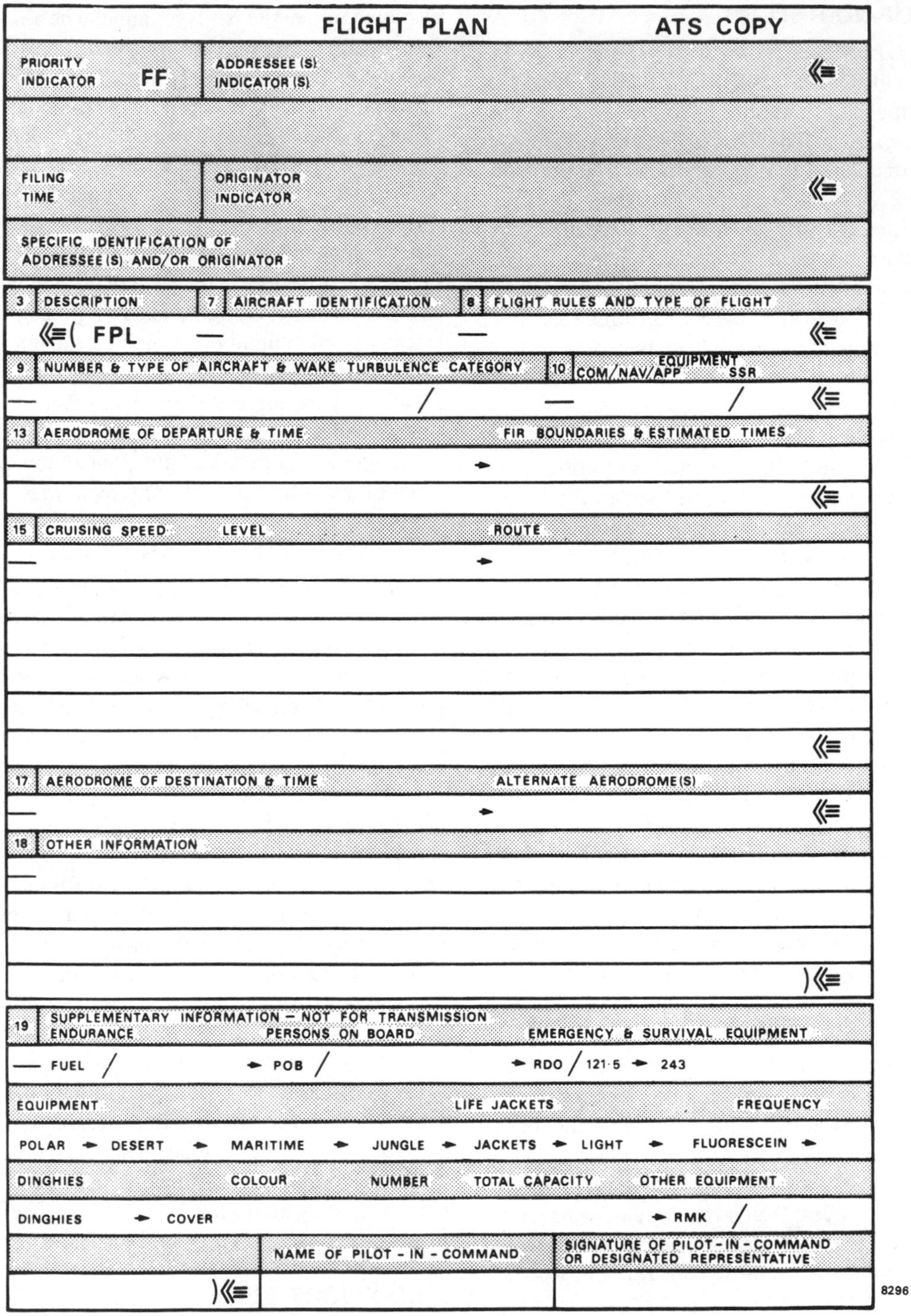

FLIGHT PLAN   ATS COPY

| PRIORITY INDICATOR FF | ADDRESSEE (S) INDICATOR (S) | | ≪ |

| FILING TIME | ORIGINATOR INDICATOR | | ≪ |

SPECIFIC IDENTIFICATION OF ADDRESSEE (S) AND/OR ORIGINATOR

3 DESCRIPTION	7 AIRCRAFT IDENTIFICATION	8 FLIGHT RULES AND TYPE OF FLIGHT
≪(FPL —	—	—	≪

9 NUMBER & TYPE OF AIRCRAFT & WAKE TURBULENCE CATEGORY	10 COM/NAV/APP EQUIPMENT SSR
— /	— /	≪

13 AERODROME OF DEPARTURE & TIME	FIR BOUNDARIES & ESTIMATED TIMES
— ➤	
	≪

15 CRUISING SPEED LEVEL	ROUTE
— ➤	
	≪

17 AERODROME OF DESTINATION & TIME	ALTERNATE AERODROME(S)
— ➤	≪

18 OTHER INFORMATION
—
)≪

19 SUPPLEMENTARY INFORMATION — NOT FOR TRANSMISSION

ENDURANCE	PERSONS ON BOARD	EMERGENCY & SURVIVAL EQUIPMENT
— FUEL /	➤ POB /	➤ RDO / 121·5 ➤ 243

EQUIPMENT	LIFE JACKETS	FREQUENCY
POLAR ➤ DESERT ➤ MARITIME ➤ JUNGLE ➤	JACKETS ➤ LIGHT ➤	FLUORESCEIN ➤

DINGHIES	COLOUR	NUMBER	TOTAL CAPACITY	OTHER EQUIPMENT
DINGHIES ➤ COVER				➤ RMK /

	NAME OF PILOT – IN – COMMAND	SIGNATURE OF PILOT – IN – COMMAND OR DESIGNATED REPRESENTATIVE
)≪		

8296

Fig. 13-5. An ICAO flight plan is required for any flight over international waters. Instructions on its completion can be found in the *International Flight Information Manual*.

pilots are familiar with is an FAA document for use in the U.S. For most flights outside of the U.S., and for any flight over international waters, an ICAO flight plan must be filed.

The ICAO flight plan is similar to an FAA flight plan, but it is much more comprehensive, and flight data must be entered on the flight plan in coded form. A copy of the ICAO flight plan form is illustrated in Fig. 13-5. You will notice that in addition to the usual information about aircraft identification and type, route of flight, altitude, and so on, blocks are also included on the form for navigational equipment, destination ETA, dinghies (rafts), and survival gear. All of this information must be entered in the proper format, using the assigned codes, and far enough in advance to allow for distribution throughout the region of flight—several hours in most cases. There is a section in the IFIM on the ICAO flight plan, including a complete description of required items and codes.

Many countries charge a fee for the filing of flight plans; others require that the flight plan be filed in person at a specific location. These requirements should be researched prior to departure from the U.S. as a part of the preliminary flight planning. (The service organizations mentioned earlier can be a great help in this area.)

EN ROUTE PROCEDURES

En route over-water procedures vary depending upon the segment of flight: coast-out, intermediate, and coast-in.

Coast-Out

The coast-out segment is that part of an over-water flight that remains within range of short range navigational aids such as VOR and DME, after departing the coast. It is during this segment of the trip that the satisfactory operation of long range navaids (LORAN-C, OMEGA/VLF, INS) should be tested and confirmed, and initial parameters for ground speed and wind correction angle established. These parameters should be compared to flight planned estimates, and if a difference significant enough to possibly affect the successful outcome of

the flight is discovered at this point, a return to the origin should be considered before any further time and fuel is wasted.

These initial estimates of ground speed and wind correction angle are also important as benchmarks for reasonableness checks and for dead reckoning back-up once the flight is dependent exclusively on long-range systems for navigation. Initial VOR/DME-based estimates of ground speed and wind correction angle can be quite accurate when done with care.

Intermediate Segment

The intermediate segment begins at that point where all short range navaids have been lost, generally between 100 and 200 nautical miles after coast-out, and ends generally the same distance from coast-in. The intermediate segment is the heart of the over-water crossing, and is worth looking at in some detail.

Navigation. Long-range, over-water navigation is not at all difficult if a modern electronic long-range area navigation system has been installed (i.e., LORAN-C, OMEGA/VLF, INS, and eventually NAVSTAR/GPS). These systems automatically determine Great Circle routes, calculate ground speeds, report present position, and provide numerous additional bits of navigational information such as distance remaining, ETA, winds aloft, drift angle, and so on. With one of these systems, long-range navigation is a simple matter of adjusting heading to maintain the desired track while monitoring the other important nav parameters: present position, ground speed, time remaining, and distance remaining.

Despite the fact that modern long-range nav systems are highly reliable, the importance of maintaining a good flight log with a record of wind correction angles, ground speeds, fuel flows, and current fuel, time, and distance remaining estimates cannot be over-emphasized. All long-range systems have limitations—LORAN-C coverage is variable and not worldwide (it does not even cover the entire North Atlantic), OMEGA/VLF is vulnerable to lane-skipping, and INS systems drift, or lose accuracy, over time. The prudent pilot always allows for the possibility of malfunction or complete failure in any case. Without a good flight log record of ground speeds, wind correction angles, and last known position, accurate dead reckoning is impossible, and accuracy in dead reckoning may be what makes the difference between a routine re-intercept of short range navaids at coast-in, and ditching.

Fuel Management. Fuel management is probably the single most important over-water en route activity. The calculation and monitoring of fuel reserves as the flight progresses—what military navigators call *range control*—is critical to over-water safety.

There are two basic methods for monitoring fuel reserves, one used predominately by civilian pilots and one used predominately by military pilots. Both are worth considering.

The civilian method parallels the conventional way to monitor fuel reserves: Estimated Fuel Remaining (EFR) over each en route fix is entered on the flight log, in an appropriately labeled column, for each fix on the flight log. Actual Fuel Remaining (AFR), normally the next column after the EFR column, is left blank for entry over each fix. (The military style flight log illustrated, Fig. 13-3, does not include columns for EFR and AFR, but the column marked A or B could be used for that purpose.)

Actual Fuel Remaining entries smaller than Estimated Fuel Remaining computations imply diminishing fuel reserves, and increasing AFR entries (compared to calculated EFR figures) imply improving reserves. When fuel reserves are diminishing, new fuel remaining estimates should be calculated for each of the remaining fixes, using revised fuel flows or ground speeds (one of which must be different to account for the discrepancy), and if fuel reserves appear to be approaching unacceptable levels, action must be taken to conserve fuel. (A revised clearance must be obtained for any significant change in airspeed and for any altitude change, but except over heavily traveled airline routes, these are usually not too hard to obtain.)

The military method uses a range-control graph to monitor fuel reserves as a function of time throughout the flight. This method is a little more complicated than the civilian method, but it is not difficult to develop or use, and it has the advantage of providing a continuous record of actual fuel remaining,

planned fuel remaining, and minimum acceptable fuel remaining at all times during the flight. Perhaps most importantly, it does so in a graphic manner that is easy to see and interpret.

An example of a range-control graph based on the fuel planning worksheet in Fig. 13-4, and the flight log in Fig. 13-3, is shown in Fig. 13-6. The vertical axis represents Fuel Remaining and has been labeled in fuel quantity increments (thousands of pounds in this case), and the horizontal axis has been labeled across the top in hourly increments. The planned fuel line is determined by plotting estimated fuel remaining versus estimated time remaining for each of the segments on the fuel planning worksheet: Level-Off plus up to four cruise segments. In this example, fuel remaining at Level-Off is estimated to be 32.5 (1,000) pounds, and time remaining at L/O is estimated to be 2 + 32 (Fig. 13-4, Climb segment). This is the first plot on the planned fuel remaining line. Each of the other planned estimates is plotted in the same way; a line between plots connects them together.

The minimum fuel remaining line is determined by plotting minimum reserve fuel (Item 13 on the fuel planning worksheet) on the zero time remaining line (bottom, far right). This point represents the minimum amount of fuel the pilot would like to see in the tanks over the destination—zero time remaining to the initial approach fix. The difference between minimum and planned plots at the zero time line is then used to plot corresponding minimum plots below each of the planned fuel remaining plots—the difference is carried down, in other words, for each planned plot—and these points are connected to create the minimum fuel line.

While en route, actual fuel remaining is plotted on this graph at various times using estimated time remaining and actual fuel remaining at that point. The difference between planned fuel remaining and actual fuel remaining can then be readily seen, as can the increasing or decreasing margin between actual and minimum fuel. (For the sake of clarity, only one actual fuel remaining plot has been shown on this graph, the circled plot at 1925 Zulu, which, as can be seen below under Flight Progress Data, is the plot for time and fuel remaining after Level Off, or what

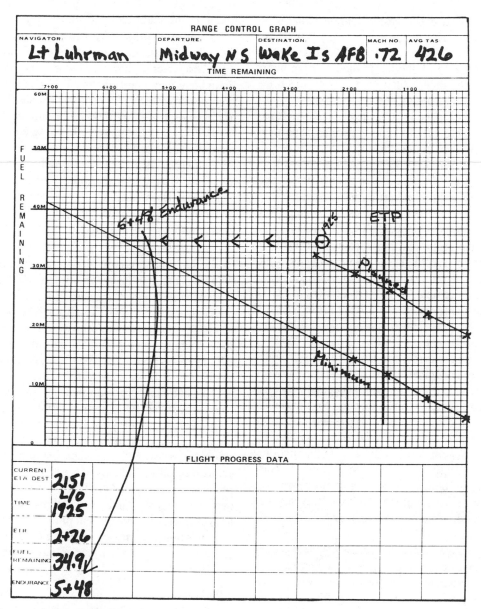

Fig. 13-6. Range control graphs provide a clear preflight picture of the relationship between planned fuel and minimum fuel; by plotting actual fuel remaining on the graph at various times during the flight, the changing relationship between actual fuel remaining and minimum fuel will also be clear.

civilians would call Top-of-Climb.)

When plotting the minimum fuel remaining line, it should be extrapolated out to the end of the graph; in this way, *endurance* can also be determined graphically, for any given time. (Endurance is the amount of time an aircraft can continue to stay airborne without having to use any of its minimum reserve fuel.) Endurance is determined in flight by first plotting actual time and fuel remaining for that point and then drawing a horizontal line from that plot until it intersects the minimum fuel line. Endurance at that point is the time remaining at the point of intersection (read off the top scale). In this case, endurance after level off can be seen to be 5 + 48; since only 2 + 26 remains in the flight at this point, this means that this aircraft is capable of staying in the air three hours and 22 minutes longer than necessary to reach the destination, not counting the reserve fuel.

Endurance can be recalculated as many times as necessary during a flight. This capability is particularly helpful when the destination weather is deteriorating: the difference between endurance and time remaining to the destination represents time available to either hold or divert to an alternate.

Note that the ETP (83 minutes remaining, from the Fuel Planning worksheet, Fig. 13-4) has been plotted on the graph and a line drawn vertically at that point to intersect both the planned and the minimum lines. The ETP is that point in the flight where range is most critical; an airport is as far away as it can be at that point in the flight (not in terms of actual distance, but in terms of time and fuel), and this is an important point to have plotted.

The use of a graph for range control provides unmistakable evidence of the relationship between planned, actual, and minimum acceptable fuel remaining, and is highly recommended for those willing to take the time to see how it works. Those accustomed to logging estimated and actual fuel remaining figures for each fix directly on the flight log can certainly continue to do so, and can, in fact, combine the best of both worlds by using those estimates to plot a range control graph. The exact way in which you keep track of fuel remaining is not nearly so important as that you do so.

Fuel management also includes the management of fuel among the various tanks. As a general rule, any fuel in auxiliary tanks should be used as soon as possible after level off, and any fuel in lockers should be transferred to main tanks as soon as sufficient room exists in those tanks to accommodate it. This procedure will normally ensure that a return to base on the main tanks can be accomplished if, for some reason, fuel will not draw from the auxiliary tanks, or fuel in the lockers will not transfer. Auxiliary fuel tanks and fuel lockers can malfunction for any number of reasons, and the fuel in those tanks should not be left unused until insufficient fuel re-

mains in the main tanks to return safely.

Communications. Position reporting is essential in non-radar environments, and position reporting over water is almost always done with High Frequency communications (HF). High Frequency is commonly called "Short Wave", and anyone who has ever used a short wave radio knows how variable HF signals can be, and what long antennas they require. There are substitutes for the long trailing wire antennas of previous eras, and frequencies are allocated so as to take advantage of the propagation characteristics of different frequencies at different times of the day, season, and solar cycle, but communication by HF is still more art than science.

Unfortunately, there are very few over-water routes that do not require HF communications capability and that can be flown without HF equipment (the Bahamas and Caribbean can still be flown primarily using only VHF). It used to be possible for general aviation pilots to relay position reports over VHF frequencies to high level turbine aircraft with HF capability, but those days are gone, except for emergencies. In an emergency, 121.5 MHz can be used to communicate with nearby aircraft who are able to establish HF communications and who can relay your position report, but those who abuse this capability and who count on being able to relay position reports over 121.5 can expect to be treated unsympathetically, both by the pilots of the aircraft required to accommodate them, and by the authorities upon arrival. General aviation pilots planning long over-water crossings will probably want to spend as much time and money researching and installing HF gear as they do long-range nav equipment.

Coast-In

The last segment is the coast-in, or landfall segment. This is the point where short-range navaids can again be received after a crossing, VHF communications are again possible, and radar contact is normally re-established. At this point, landfall can be calculated with some assurance and accuracy, and a return to relatively normal navigation and communication anticipated.

Initial radar contact is the time of reckoning, both

for the long-range system, and for the pilot, since this is the first point where drift and cross-track error can be positively identified. Aircraft found exceeding certain route parameters—20 nautical miles in many cases—can expect a report to be filed. In some cases, an explanation will be demanded, and fines are occasionally levied. This should be no concern for pilots with properly functioning long-range systems who monitor their track carefully and who observe all over-water regulations regarding in-flight contingencies (printed on the appropriate Jeppesen charts, and included in the IFIM), but it is important that pilots understand that over-water navigation also means over-water separation, and that every effort be made to stay as close to cleared routes, altitudes, and airspeeds as possible, for their own sakes, if not for others.

CONCLUSION

Over-water navigation is more complicated than land navigation, because fewer sources of information are available, errors are harder to detect, and the consequences of errors are more critical. This does not mean that over-water navigation is especially difficult or dangerous, but it does mean that equipment must be in good working order, that extensive and careful preflight planning is absolutely necessary, and that the pilot must be proficient in all aspects of air navigation from dead reckoning to long-range area navigation. With a conservative and methodical approach to flight planning, and with careful attention to detail, over-water crossings can be routine and uneventful, increasing the utility of appropriately equipped aircraft significantly.

Chapter 14

Navigation In Non-Radar Environments

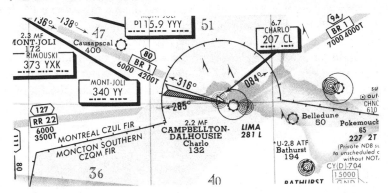

From a purely technical point of view, this chapter is not so much about navigation as it is about Air Traffic Control (ATC). Ground based radar (such as that used by ATC) can be used to facilitate any form of navigation, but its presence or absence does not change the navigational system in use. Navigation in non-radar environments is not a different form of navigation—it is simply a different operational environment. Nonetheless, the procedural differences between radar and non-radar environments are sufficiently great that navigation in non-radar environments seems like a different form of navigation. Accordingly, this chapter will look at those procedures necessary to operate safely in non-radar environments, treating the non-radar environment as a special navigational case.

BACKGROUND

In the United States, radar monitoring of aircraft on instrument flight plans is so common that pilots tend to think of the radar environment as the norm, and the non-radar environment as the exception. In fact, it is the other way around: all air traffic control is based, fundamentally, on non-radar procedures. Radar is an addition to the basic system—an addition that facilitates air traffic control enormously and reduces pilot workload substantially—but an addition that is not, in itself, necessary for the system to work.

Because most pilots operating in the U.S. have come to assume that radar control will be available, they are often taken by surprise when it is not. They shouldn't be. When radar contact is lost, the pilot is forced to revert to non-radar procedures, but these are actually the normal procedures; radar simply temporarily suspends their necessity.

NON-RADAR AREAS

Most of the world, in fact, is not under radar control or surveillance. This is sometimes difficult for U.S.-based pilots to accept, since so much of the United States is blanketed by radar coverage, but nonetheless, most navigation, over most of the world, is done without the benefit of ground-based radar guidance or supervision. Most of Canada north of a line extending from Quebec City through Montreal, Ottawa, and Toronto, for instance, is not under radar control. A good part of the Bahamas and Caribbean is outside of radar control. Most of Central and South America, Africa, Asia, and the Pacific are without radar coverage, and, except for coastal areas, there is no radar coverage over water. Even on a relatively short over-water route such as New York to Bermuda, over a third of that distance is outside of radar control.

In the United States itself, radar coverage can be lost due to computer failures in the air traffic control facilities, and ground-based radar can itself fail for any number of reasons—radars are very complex pieces of equipment, requiring high power and very finely tuned individual components to function properly. In addition, since radar operates in the microwave frequency band, all radar transmissions are line-of-sight. This means that radar coverage will be lost in hilly or mountainous terrain below certain altitudes due to terrain masking, and radar coverage will be lost below a certain altitude due to curvature of the earth in any area. Airborne transponders decrease the masking effects of the terrain to a certain extent, but not entirely.

Pilots can therefore expect to find non-radar areas at all times and at all altitudes in many parts of the world, and at some times and some altitudes in every part of the world, including the United States. Since radar coverage is not an absolute, but varies with terrain, altitude, condition of equipment, and atmospheric conditions, radar coverage charts are not possible. The pilot who assumes that there is always a possibility that radar coverage will not be available will be in much better shape than the pilot who always assumes that radar coverage will be available.

POSITION REPORTING

When radar is not available, the only way ATC can control air traffic is via position reporting—the relay of position information from pilot to ground controller. This means that in a non-radar environment the pilot must be aware of his position at all times, he must know when a report of that position is required, and he must know how to make a position report that is accurate and complete.

The identification of present position is the es-

sence of navigation itself and the subject of a good part of the rest of this book; therefore, while we will look at ways to verify present position in this chapter, the specific techniques for determining present position are not our concern here. Our concern here is knowing when to report, and how.

Terminology

The pilot should assume that he is not in radar contact unless he is specifically told otherwise ("Radar contact"). Once in radar contact he can assume that he continues to be in radar contact until he hears the phrase "Radar contact lost" or "Radar service terminated". These are the three key phrases that indicate to the pilot whether position reporting will be required or not.

Reporting Points

In order to facilitate reporting and control, ATC has designated and marked those points, or fixes, where a position report is either essential or likely. Essential reporting points are called *Compulsory Reporting Points*, and fixes that are likely to be used for reporting points are called *On Request Reporting Points*. Compulsory Reporting Points are marked (on the instrument navigation charts—both Low and High Altitude) by solid triangles; On Request Reporting Points are indicated by open triangles. (See Fig. 14-1.) For aircraft cleared along direct routes (i.e., off airways), the pilot must also report each fix to which he is directly cleared—a direct fix is automatically a Compulsory Reporting Point.

ATC can request that a position report be made anywhere—every fix or position, including arbitrary DME fixes or crossing radials, is a potential On Request Reporting Point. The pilot should therefore not assume that position reports will only be required over designated fixes. In practice, however, there are so many fixes already plotted and marked as either Compulsory or On Request Reporting Points, that the need for a report over an unmarked fix is rare.

THE POSITION REPORT

The position report itself follows a specific format designed to provide the controller with all the

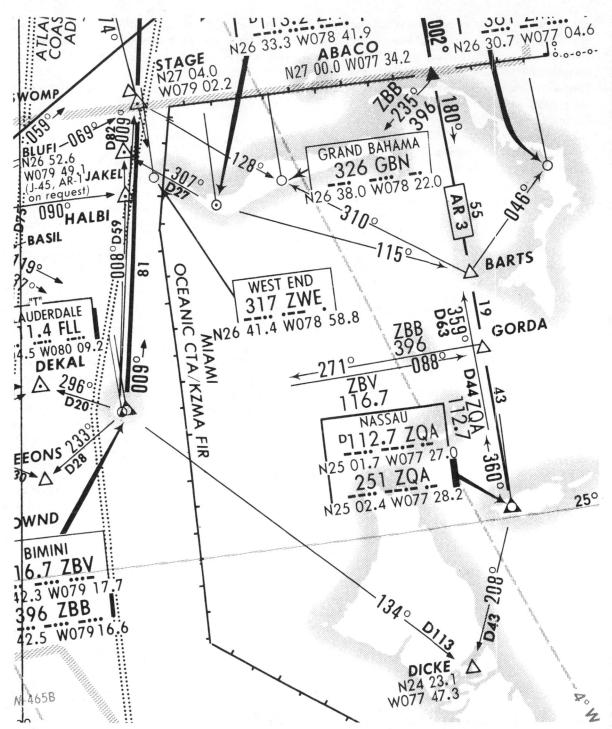

Fig. 14-1. NASSAU and ABACO are Compulsory Reporting Points. DICKE, GORDA, and BARTS are On Request Reporting Points. Reproduced with permission of Jeppesen Sanderson, Inc. Not actual size. Not to be used for navigation.

information he needs to plot position and control separation without the need for further questions. This makes the position report format seem unnecessarily complicated to pilots who are not used to making them, but perfectly logical and concise to those who are. The position report format is:

1. Aircraft Identification.
2. Position.
3. Time.
4. Altitude or Flight Level.
5. Type of flight plan. (Not required for IFR reports made directly to ATC.)
6. ETA and name of next reporting point.
7. Name of the next succeeding reporting point.
8. Any remarks.

Identification. The aircraft identification comes first so that the controller can immediately zero in on the appropriate flight strip for the remainder of the report.

Position. After the aircraft ident itself, the next most important piece of information is the aircraft position. The position being reported will be either a Compulsory Reporting Point, an On Request Reporting Point that has been requested, or any fix to which the pilot has been cleared directly.

Time. The time (to the nearest minute) over the fix being reported is always included, even if the time is "right now." There are several reasons for this. In most cases it takes a minute or two to work up an accurate position report. It also sometimes takes time to establish contact with the controller. In some cases the report will actually be relayed to a controller, either by FSS or by *ARINC*. (ARINC, Aeronautical Radio, Inc., is a radio relay service owned by the airlines and available, on a fee basis, to all pilots.) The relay itself can take several minutes, depending on the means of transmission and the workload of the radio operators and controllers involved. Each of these situations increases the time between the actual arrival of the aircraft over the fix and the time the report is received, and explains why time must be included in the position report.

Time is always Coordinated Universal Time (UTC), what we used to call Greenwich Mean Time (GMT) and what is usually called Zulu time in aviation. (The English no longer keep track of the time at Greenwich, the French do, in Paris, which is why the initials, UTC, don't match the order of the words—the abbreviation is from the French.) Since it is easy to make mistakes converting local time to Zulu time, especially when changing time zones, it is a good idea to leave one clock in the cockpit set to Zulu time at all times (and it is also a good idea that it always be the same clock). Then, when a position report is required, the additional complication of having to remember or look up the necessary conversion factor is eliminated. Zulu time is the same all over the world, so, once set, this clock can be left alone regardless of changes in time zones, and regardless of seasonal changes from Standard to Daylight time.

In order to achieve the greatest accuracy and to insure consistency among aircraft, the FAA is very specific about how to establish the exact time over the fix. If the fix is a VOR facility, the time noted should be the time at which the first complete reversal of the To/From indicator is accomplished; i.e., when the indicator flips from Off to From. Likewise, if the fix is an NDB, the time noted should be that time when the needle has made a complete reversal. For fixes that cover a broader area than a VOR or NDB, such as a marker beacon, the time when the signal is first received should be noted, as well as when it is last received, and the time over the fix estimated as halfway between these two times. When Lines of Position or DME fixes are used, the FAA simply says that "the distance and direction should be computed as accurately as possible." (*Airman's Information Manual*, paragraph 341. [4].).

Some pilots mistakenly assume that since control of aircraft by non-radar methods isn't very accurate to begin with that there is little point in worrying about accuracy in noting times. This is, in fact, exactly wrong. Because non-radar methods aren't very accurate to begin with, every effort should be made to not introduce additional errors; therefore, accurate and consistent timing of position over reporting points is very important.

Altitude or Flight Level. Altitude is the vertical component of position. The controller knows what your altitude is supposed to be from your clearance; this is a double check on that information.

Type of Flight Plan. When a facility such as FSS or ARINC relays a position report, it is necessary to include the type of flight plan: IFR or VFR. This is to ensure that IFR position reports are relayed to the appropriate ATC facility. (VFR position reports are simply filed for reference in case the reporting aircraft becomes overdue.)

ETA and Next Reporting Point. The normal format for estimates is: "Estimating XYZ at 1234 Zulu." Simply stating the ETA and then the fix name—"1234 Zulu, XYZ"—is technically correct, but spelling it out in plain English helps eliminate any confusion as to which time goes with which fix, and only adds a few words to the transmission.

Name of Succeeding Reporting Point. The name only of the next succeeding reporting point—the fix after the next fix—is included at the end of the report as a further verification of the route and direction of flight. Like the altitude report, it is not absolutely essential, but it serves as an important double check on the clearance.

Remarks. The remarks section is an appropriate place to relay pilot reports of winds aloft, temperatures, cloud tops and bases, icing, and any unforecast or hazardous weather conditions, and ride reports. In many parts of the world pilot reports of actual meteorological conditions are the only source of weather information available and pilots are encouraged to include pilot reports of flight conditions at the end of their position reports, and in some cases are required to. Over the North Atlantic, for instance, air carriers must include the winds aloft and temperature in the remarks section, and all others are requested to do so.

Pilots sometimes invent memory aids to help in remembering the position report format; however, if the purpose of the position report is kept in mind—to accurately relay to the controller the aircraft's position (ident, fix, time, and altitude), provide an estimate to the next fix, and indicate the fix that follows—there should be little trouble in remembering what is required: "November 1234 Xray reporting St Maarten at 1723 Zulu, FL 120, IFR, estimating Elopo at 1734 Zulu, Coolidge." (This example is based on actual reporting points along Blue 520. See Fig. 14-2.)

ESTIMATING TIMES

In the absence of radar, the controller cannot, of course, actually see the aircraft under his control, but must visualize their movements based solely on aircraft reports of position and estimates of their arrival at the next reporting point. One of the assumptions upon which this system is based is that all estimates of time will be accurate to within three minutes, and that the True Airspeed, as provided on the initial flight plan or as amended in flight, will be accurate to within 10 knots or five percent of actual TAS, whichever is greater. In practice, most modern aircraft are equipped either with airspeed indicators capable of providing TAS information accurate to within 10 knots, or Mach compensators and air data systems capable of even greater accuracy. Therefore the allowance of five percent for aircraft faster than 200 knots is seldom necessary (five percent of 400 knots would be 20 knots, for instance) and it is generally assumed that reported TAS will be accurate to within 10 knots for all aircraft.

The controller takes these two pieces of information—estimate to the next fix and TAS—along with similar information from other traffic under his control, and determines an appropriate plan to ensure aircraft separation. Where position reports and estimates indicate that aircraft separation standards will not be met, the controller will issue a revised clearance to one or more aircraft to provide additional separation. These amended clearances will generally take the form of airspeed changes, new routings, or holding instructions.

The controller assumes, unless he is told otherwise, that the aircraft will actually arrive at the next fix within plus or minus three minutes of the estimate given, and has predicated his plan of control on that basis—there is no additional margin for error. The closer the pilot comes to arriving exactly at his estimated time, the greater the separation. To ensure minimum safe separation, all time estimates must be accurate to within three minutes, and reported TAS must be accurate to 10 knots or five percent.

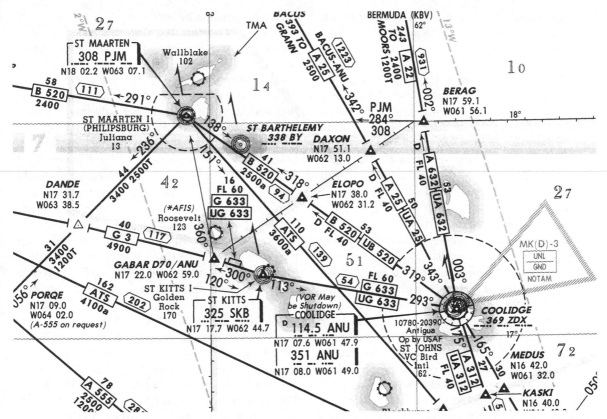

Fig. 14-2. An IFR position report over ST MAARTEN en route to Antigua (COOLIDGE) on B 520 would include aircraft ident, the name of the fix (ST MAARTEN), the time over the fix, the aircraft altitude, IFR, the name and an estimate for the next fix (ELOPO), the name of the fix after that (COOLIDGE), and any remarks. Reproduced with permission of Jeppesen Sanderson, Inc. Not actual size. Not to be used for navigation.

DME Available

When DME is available, estimating time to the next fix is a fairly simple matter. Once the ground speed readout has stabilized at cruise, estimating time to the DME fix in use is as easy as flipping the switch to the Minutes position, and then adding minutes indicated to present Zulu time. This will generally result in an estimate accurate to at least three minutes. (At extreme DME ranges the error may be greater than plus or minus three minutes; however, as the aircraft gets closer to the station the error will become apparent and a revised estimate can be forwarded.) If the DME installed in the aircraft does not have a Minutes position, then minutes to the station will have to be computed with a DR computer using the distance readout and the indicated ground speed.

DME Not Available

DME provides very accurate ground speed estimates (and therefore time estimates), if used properly. In a non-radar environment however, DME ground facilities are not always available. In fact, if radar is not available, all too often DME is not available either. In these cases, unless some other navigational system such as LORAN-C or OMEGA/VLF is available, the pilot must rely solely on dead reckoning for his estimates.

Dead reckoning is only as accurate as the facilities in use and the care with which the pilot makes his computations. Certain types of facilities are more accurate than others (VOR is more accurate than NDB, for instance), distance from the station is an important factor in accuracy, and crossing radials or bearings (Lines of Position) that are 90 degrees to each other are more accurate than LOPs at oblique angles to each other.

Accurate tracking and precise station passages are also very important in dead reckoning: the more accurate the tracking, the more closely the distance traveled will match the distance measured, and the more precise the station passage, the more accurate the time between fixes. Care in setting and reading the DR computer will also result in more accurate estimates. (Electronic calculators naturally provide the most accurate computations, but mechanical computers are perfectly adequate if used carefully.)

Estimates should be regularly re-computed to ensure that the estimate remains accurate to within three minutes. The more often ground speed is re-computed and the more often cross-checks of position are made, the more accurate the estimates will be. Generally, the most accurate ground speed estimate will be an average of the last several computations, and not the very last computation itself. There will be times, however (for instance, when the winds aloft are changing rapidly) when the last computation will be the most accurate. (A substantial change in the wind correction angle, the onset of turbulence, or an abrupt temperature change, all indicate wind shifts.) This is an area of pilot judgement, but if in doubt, a ground speed that has been averaged over a fairly large distance will usually be the most accurate.

It is important to remember that these estimates are not something you do simply for the benefit of the controller: the more accurate the estimate, the greater the separation between you and conflicting traffic. In any case, since you are dependent upon the estimates of other pilots for your own separation, greater accuracy on your part may be necessary to compensate for lesser accuracy on theirs. You are, in other words, the direct beneficiary of accurate estimates.

POSITION VERIFICATION

Radar provides an automatic double check of position—both pilot and controller monitor the aircraft movement—but in a non-radar environment, the controller has no way to verify the aircraft's reported position. It is therefore very important that the pilot verify his position himself.

Flight instructors frequently teach their primary students to have at least two, and preferably three, positive ground-based references to be certain of their position when navigating via pilotage and dead reckoning: many small towns have rail lines passing through them, for instance, but only one, presumably, will have a railroad track running northwest to southeast with a major highway crossing it near a tall tower. Likewise, the pilot of an aircraft operating in IMC (Instrument Meteorological Conditions) should have at least two and preferably three positive electronic references to verify his position—the wrong VOR will indicate station passage just as positively as the correct one will, and without an additional source of information there is no way for the pilot to know that the wrong VOR was used. The only way to know for sure where you are is to use all the available cockpit resources to identify and verify position: VOR, DME, NDB, LORAN or other long range system, and visual confirmation (pilotage) when available. Intersections (as opposed to navigational facilities themselves) should also be identified in as many different ways as possible: additional VOR radials, DME information, NDB bearings, and supplementary data from LORAN, OMEGA, INERTIAL or other long range area systems.

Time should be used to check for reasonableness. Early or late station passages should be suspected and double checked. If present position turns out to be correct, then that means that the previous ground speed estimate was in error and ground speed should be re-computed.

If you are where you think you are, then every piece of evidence will agree (although not exactly—some of the evidence will be more accurate than others). Any one piece of evidence to the contrary should be regarded as a warning of a possible discrepancy (but only a warning); many pieces of evidence to the contrary are a clear warning of an error. The pilot must resist both the natural tendency to see only what he wants to see and ignore the rest, and the tendency to focus on and overemphasize a single piece of conflicting information. This is where judgement and experience takes over and hard and fast rules leave off; however, as long as the pilot has a good understanding of the limitations of each type of navigational system, and uses all the information available, he should have no trouble sorting out important conflicting information from aberrations (quirks) or errors in verification.

UNCONTROLLED AIRSPACE

In uncontrolled airspace, there is no coordination of aircraft movement and no separation is provided by ground-based controllers. Pilots are free to operate in the clouds at their own discretion, so long as they maintain at least 1,000 feet of clearance above all obstructions within a five-mile radius (2,000 in areas designated as "mountainous terrain"), and so long as the IFR hemispherical rule is observed (FAR 91.121 (b).).

To operate in the clouds without any form of air traffic control may seem like the height of folly, but in fact it is not uncommon in isolated parts of the world to find not only large areas of uncontrolled airspace regularly being used by local traffic, but to find uncontrolled areas served even by the regularly scheduled airlines. (Figure 14-3 shows a large area of uncontrolled airspace not too far north of Presque Isle, Maine, that is served daily by regional airlines using Transport Category equipment; the area includes several airports that are popular arrival points for fishing expeditions.) Pilots operating in uncontrolled airspace are expected (but not required) to report their position "in the clear"—i.e., to whomever is listening—and to coordinate their own separation as necessary. Surprisingly enough, very few midair collisions have occurred in IMC in uncontrolled airspace, although clearly the system (or lack of it) has its limits.

Pilots should at least be aware that such a thing as uncontrolled instrument airspace exists, that it is not as uncommon or perhaps as dangerous as they might think, and that position reporting is especially important in uncontrolled airspace, even though it is not required. Position reporting in the clear is the only way airspace conflicts can be resolved.

TRANSITIONS

The transition from a non-radar to a radar en-

vironment is straightforward: The controller normally asks for an ident, and once the aircraft has been positively identified, he then informs the pilot that he is "Radar Contact". Controllers outside of the U.S. who follow ICAO (International Civil Aviation Organization) rules will add "Discontinue further position reports," or words to that effect. The pilot is free of the requirement to make position reports from that point on.

The transition from a radar to a non-radar environment is only slightly more complicated. The controller will advise "Radar Service Terminated" or "Radar Contact Lost." In most cases he will also advise the pilot at the same time of the requirement for a position report at the next Compulsory Reporting Point, and advise the pilot to whom that position report should be given. Even if he doesn't, the pilot should immediately begin working on an estimate to the next fix once radar contact has been lost—the controller can require an estimate to the next reporting point at any time, and the pilot is expected to have that estimate available when asked. Whether requested or not, over the first mandatory reporting point the pilot should make his report—ident, fix, time, altitude, [IFR], estimate to next fix, and fix after—and continue making position reports until advised that he is again in radar contact.

SUMMARY

In many parts of the world, the non-radar environment is the norm. In every part of the world, non-radar procedures form the basis for all aircraft control. Navigation in a non-radar environment is no different from navigation in a radar environment, except that in a non-radar environment more emphasis is placed on pilot awareness of position, and accurate ground speed information is essential for accurate estimates. The position report itself consists of the minimum amount of information necessary to completely describe the aircraft position and anticipated route of flight: aircraft identification, position and time, altitude, type of flight plan, an estimate to the next fix, and the name of the fix to follow that one.

Navigation in a non-radar environment is a little more demanding than navigation in a radar en-

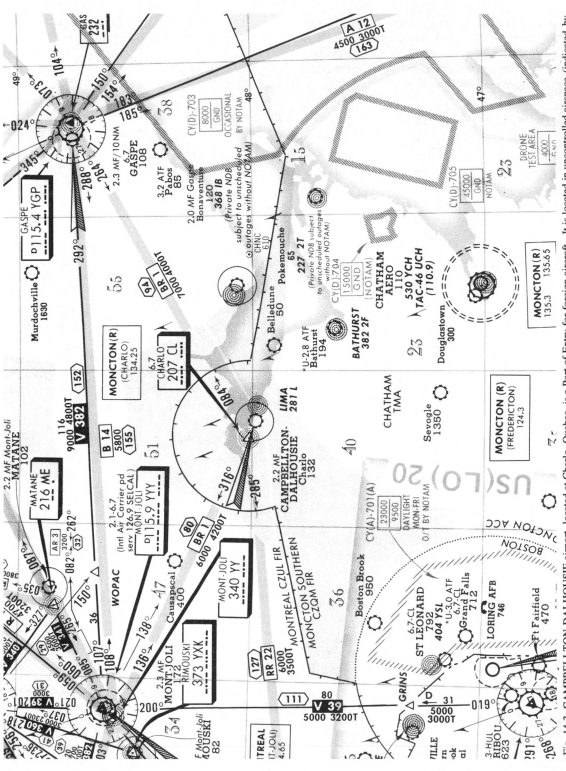

Fig. 14-3. CAMPBELLTON-DALHOUSIE airport, located at Charlo, Quebec, is a Port of Entry for foreign aircraft. It is located in uncontrolled airspace (indicated by tinted shading) and is served by regional airlines. Reproduced with permission of Jeppesen Sanderson, Inc. Not actual size. Not to be used for navigation.

vironment in that the non-radar environment requires continuous monitoring of position and ground speed and the need for frequent position reports. But for the pilot who understands his navigational systems and who is in the habit of keeping an accurate record of times and estimates (a flight log), navigation in non-radar environments should present few problems.

Chapter 15

Radar Terrain Mapping

The military has long used airborne radar for primary navigation, however airborne radar has never been used for primary navigation in civilian aviation. This is as it should be: the military has to rely on radar for navigation and terrain avoidance—there is only one other navigational system suitable for all-weather, high-speed, low-altitude operations—infrared—and infrared is still very new and somewhat experimental. The military has radar systems that scan in 360 degrees, displaying a ground plan of the prevailing terrain, and they also have radars which project a profile view of the terrain ahead, enabling the pilot to see the low points between ridges for even greater accuracy in high speed, very low altitude runs in all kinds of weather. Radar guided ''nap-of-the-earth'' flying is an inherently high-risk activity though, one that is necessitated by tactical requirements; this kind of navigation is obviously totally inappropriate for civilian operations.

This is not to say that airborne radar navigation does not have its place in civilian aviation though— the terrain mapping capability inherent in airborne weather radar can be quite useful in normal civilian operations, and it may be the only source of navigational information in an emergency. Radar terrain mapping is a very inexact science, and a considerable amount of experience is necessary to reach even minimum levels of radar proficiency, but it does have its place.

This chapter will look, briefly, at some of the ways in which airborne radar can be used to supplement normal navigation and to navigate in an emergency. This is not meant to be a comprehensive look at radar terrain mapping theory and technique, nor is it meant to be an instruction manual in radar operation: every radar set is a little different and requires different techniques for optimization. This chapter is meant to provide a starting point for incorporating radar into normal navigational routines, and it is meant to provide examples of some of the ways in which radar can be used in an emergency for primary navigation.

RADAR TERRAIN
MAPPING CHARACTERISTICS

Radar is a pictorial representation of reflected radio

energy. Normally, airborne radar is used to display relative levels of rainfall—radio energy is reflected in direct proportion to the amount of moisture present in the air. This characteristic of radio energy to be reflected by particles in its path is not limited to weather depiction, however, but can also be used to display terrain characteristics. This function is known as radar terrain mapping (also called *ground mapping*), and the associated radar CDU control for optimization of the terrain mapping function is normally labeled Map (see Fig. 15-1).

What makes terrain mapping possible, however, is not that terrain reflects radio energy, but that it reflects energy in varying amounts depending upon its physical characteristics. (If all terrain features reflected radio energy equally, distinguishing among different terrain features would be impossible.) Manmade objects, for instance, reflect a large amount of energy and produce strong radar returns; mountainous terrain also reflects a large amount of energy, but less than manmade objects do; flat terrain reflects somewhat less energy than mountainous terrain does, and water reflects hardly any energy at all. (The exact amount of radio energy reflected by water depends

on the surface texture of the water: smooth water reflects virtually no energy; choppy water reflects a small amount.) These characteristics enable the pilots of radar-equipped aircraft to distinguish between mountainous areas and flat terrain, between sea and shore, and between urban areas and the surrounding terrain.

Radar functions by transmitting energy in a narrow beam that is swept back and forth to create, in effect, a plane of radiated energy. This plane of radio energy travels in a line-of-sight pattern until it hits something and is reflected back, or fails to hit something and continues off into space. The beam can be aimed up or down by the tilt control on the radar set in order to focus the area in which the energy is directed: straight ahead (more or less) for weather detection, downward for terrain mapping. (The tilt control is the large knob on the far right in Fig. 15-1.)

For proper ground mapping, the tilt control must be adjusted carefully or terrain will be cut out of the picture: too high, and the radar beam will only hit the terrain at the far horizon (or miss it altogether—the normal weather position); too low, and the radar will show only the terrain directly in front of the aircraft. The greater the range selected, and the lower the altitude, the greater will be the tendency to cut off either the near or the far terrain—low and far away, the radar will be looking at the terrain at a flat angle, rather than from above, reducing the area scanned. With some combinations of range and altitude, it may be impossible to show the entire terrain from just under the aircraft to the horizon, and a compromise may have to be struck.

The only way to determine the proper tilt angle is by trial-and-error, the goal being a full screen of terrain information, with no blank areas at the top or bottom of the screen. The tilt must be readjusted after any significant change in altitude, since altitude changes the relative angle between aircraft and the horizon. Once a good tilt angle has been established, fine-tuning of the angle will sometimes improve the quality of the image by optimizing the angle between beam and terrain toward the perpendicular—the angle of maximum reflection.

The MAP position should always be selected whenever terrain mapping is desired (as opposed to

Fig. 15-1. The terrain mapping mode is selected with the square button marked MAP on this radar CDU. The MAP mode provides the clearest and most complete terrain picture. Copyright Sperry Corporation, 1985.

weather detection). Airborne weather radar is normally optimized to detect the very small amounts of energy returned by raindrops. The much stronger returns from terrain tend to overpower the receiver circuitry.

Also, the terrain directly under and in front of the aircraft will often be missed entirely in the Weather (WX) position, because the radiated energy returns so quickly that the radar normally doesn't have time to switch over from transmit to receive before it arrives. The Map position optimizes the receiver circuitry for terrain mapping by introducing a delay for close-in returns, and it reduces the sensitivity somewhat to compensate for the relatively stronger returns from terrain. In many cases, switching between the Map and the WX positions will not seem to produce any difference in what is displayed, but the Map position should still be selected for terrain mapping as a matter of habit—there are times when it will make a difference. (Conversely, the WX position should always be selected for weather detection or significant areas of rain may be eliminated.)

NORMAL USES

There are three main areas of normal radar terrain mapping. In descending order of importance they are: coastline and island mapping; identification of cities and towns; monitoring of mountainous areas.

Over Water

The most common and probably the most important use for radar as a terrain mapping tool is in identifying coastlines, islands, and other terrain characteristics common to the juncture of land and sea: harbors, bays, peninsulas, inlets, hooks, barrier islands and so on. Since water normally reflects almost no energy, while land reflects a considerable amount, the dividing line between land and sea is usually quite apparent on the radar screen, and this characteristic can be useful in overwater operations.

It is not uncommon, in fact, for radar to provide the first cockpit indication of the arriving landfall during an over-water crossing: while VOR, DME and radar all are limited to line-of-sight transmissions, radar is often the system with the longest reach. By

setting the range to maximum line-of-sight distance for the cruise altitude, and by adjusting the tilt toward the horizon, landfall can often be seen on the radar screen prior to re-intercept of VOR/DME information and prior to ground-based radar contact.

To determine the line-of-sight range for any given cruising altitude (and therefore to determine the maximum distance the radar can see for any given altitude) use the following formula:

$$D = 1.23 \times \sqrt{h}$$

where D is the line-of-sight distance in nautical miles, and h is the altitude in feet. To determine the line-of-sight distance from FL 390, for instance, use a calculator to find the square root of 39,000, then multiply that answer by 1.23: 243 nautical miles. (In order to take advantage of the full 300 nautical mile range capability common to many of the newer radar sets, the aircraft would have to be operated at FL 600; as far as I know, the Concorde SST is the only civilian aircraft capable of operation that high.)

Radar resolution—the ability to picture detail—is poor, and in fact, the newer, color digital sets, which process raw radar data into digital form to provide a non-fading, color graded display, have even less resolution than older, black-and-white, analog sets. (There are some pilots who still prefer the older sets for just this reason.) Since resolution is not especially good with any radar, only prominent terrain features will appear on the screen—the pilot cannot expect weather radar to show every nook and cranny along a shoreline. Without significant, individual characteristics, such as an arm or a hook extending into the sea, or a bay or large harbor breaking the shoreline, identifying specific points along the shoreline can be difficult. Islands are easier: large, distinctive islands like many of those in the Bahamas and Caribbean can be quite easily identified. The more distinctive the coastline or island shape, the easier the terrain will be to identify.

Just as islands—land areas surrounded by water—can be seen quite easily on radar, the opposite also holds true: water surrounded by land—lakes—can also be easily seen. Small lakes and ponds will not appear (the exact size where lakes disappear depends mainly on the range setting), but large lakes will show up nicely, and this characteristic can be very useful in instrument conditions.

Water that is choppy or rough will reflect some energy (since the near sides of the waves provide a good reflecting surface) and will often look like areas of light rain. This effect can sometimes be negated, or at least minimized, by adjusting the tilt upwards, creating a more oblique angle between the radar beam and the waves and reducing the amount of energy reflected. It can also sometimes be eliminated by tilting the antenna downwards, since that directs the energy down into the bottoms of the waves and away from the sides. Even with some reflection off the water though, the shoreline can usually still be picked out, since more energy will be reflected off the land than off the waves; waves reduce the resolution and blur the image, but seldom eliminate the break between shore and water entirely.

Ice is similar to water in that smooth ice reflects very little energy, while rough ice reflects quite a bit. Rough ice (Arctic ice is often rough, for instance) looks like land, and the radar is of little value in these areas.

Snow tends to absorb radar energy, which reduces its reflectivity. Snow covered bodies of water will therefore often look like snow covered areas of land; however, if the shoreline rises up above the water in the form of cliffs, dunes, or hills, then it may still be possible to make out the shoreline, since the strongest returns come from beams that strike terrain at perpendicular angles. Adjusting the tilt to fine-tune the angle between beam and terrain will sometimes bring out a shoreline that otherwise cannot be seen.

Cities and Towns

Cities and towns stand out quite clearly on airborne weather radar. Steel and concrete provide very strong returns, stronger even than mountains, and any city or town with tall buildings, bridges, factories or other substantial manmade structures will appear brightly on the screen, even if surrounded by hilly terrain. (Naturally, if the surrounding terrain is high enough and the radar angle is low enough so that the terrain obscures the manmade features, then they will

not appear.)

Unfortunately, airports do not usually provide distinctive returns; the large amounts of concrete common to airports are still too small to be resolved apart from the weaker but still fairly strong returns produced by the surrounding terrain, and the angle between beam and terrain is less than optimum—flat instead of perpendicular. It is, however, often possible to pinpoint the location of an airport on the radar screen if its position relative to the strong return from a nearby city is known.

Mountainous Terrain

The near sides of mountains provide excellent reflecting surfaces, and mountainous areas, ridges, and even individual peaks can often be seen on the radar. Civilian weather radar does not have the resolution or computerized sophistication to plot routes or identify specific locations based solely on terrain returns, but it can still be used for orientation relative to mountainous areas.

The flatter the tilt angle, the better the leading edges of the mountains and ridges will stand out (just as bright sunlight from a low angle makes details stand out sharply late in the day, while the even lighting of an overcast day eliminates all surface detail). Mountainous terrain therefore shows best when scanned at long ranges and from relatively low altitudes, raking the area with as flat a beam as possible. As the aircraft approaches the area of mountainous terrain, or as it gains in altitude, the mountains may seem to disappear and be replaced by flat terrain: as the aircraft gets closer to the mountains or increases in altitude over the mountains, the radar beam no longer hits the sides of the mountains, but instead radiates down into the valleys where it is reflected off all surfaces alike.

These then are the normal, back-up uses for weather radar as a navigational, terrain mapping tool: shoreline, lake, and island identification, identification and location of cities and towns, and monitoring of mountainous areas. It is important not to overestimate the ability of radar to map terrain—accuracy and reliability in civilian radar terrain mapping is not nearly good enough for primary navigation—but it is important also to be aware of its usefulness as a back-up source of information. Regular use of radar in routine operations helps prevent occasional gross errors in navigation, and leads to the knowledge and experience necessary to optimize radar's emergency capabilities when that becomes necessary (the subject of the next section).

EMERGENCY USES

Airborne radar can be used in two different ways in an emergency: it can be used to assist with pilotage, and it can be used to assist in dead reckoning. Neither technique is terribly accurate, but both are quite a bit better than nothing.

Pilotage Mode

Over-water, airborne weather radar can be used to look ahead for a landfall and proceed, in an emergency, to a fix on shore based on radar identification of prominent terrain features—electronic pilotage. An aircraft making the normal North Atlantic crossing from west to east, for instance, would expect to arrive at one of several coast-in fixes to the west of Ireland, Scotland, England, and France. The European coastline is very distinctive, even on large scale instrument charts, as can be seen in Fig. 15-2. (Long-range visual charts, such as Operational Navigation Charts—ONCs—provide even more detail.) Most of these terrain features would be apparent on the radar screen at line-of-sight distances from shore—200 to 250 nautical miles at normal cruising flight levels. By relating the radar picture to the chart, it is a fairly simple matter to correct back onto the proper track.

Dead Reckoning Mode

Ground speed can be obtained with a DR computer (an E6-B) by measuring the elapsed time to cover a known distance; aircraft movement between fixes can be observed either visually or electronically (by measuring the elapsed time between two geographic or airway fixes, for instance). Radar can also be used to obtain a ground speed estimate. Radar-based dead reckoning will generally be less accurate than visual or VOR-based dead reckoning, but a radar fix can often be taken in instrument conditions

where a visual fix cannot, and radar can be used in an emergency when no other electronic navaids are available. Distance can be obtained directly from the range markings on the radar screen—only one target is required and the target does have to be identified to be usable.

To determine ground speed, first locate a specific geographic target at the top of screen: the center of a city area, a prominent terrain feature, a mountain peak—its actual identity is irrelevant, but the more distinctive the better, and the closer to the top of the screen the better. (*Distinctive* means smaller, but if it is too small it may be difficult to track.) Large ships can sometimes be found on the radar screen, which means that this technique can even be attempted in the middle of the ocean. (Ships are moving targets, of course, and their relative speed will distort the results, but not substantially.)

Once a target has been chosen, start timing as the target crosses the top range line. Note the elapsed time as the radar fix passes the next range mark. Use elapsed time and distance between range marks to figure ground speed in the conventional manner.

Track (and from that, wind correction angle) can also be derived from radar data by plotting the distance and azimuth to a radar fix two or more times on the wind side of the E6-B. (Azimuth to the fix can be determined using the azimuth markings on the radar screen: add relative azimuth to aircraft true heading. See Fig. 15-3.) By aligning these plots with the vertical axis of the wind slide, track can be read under the True Index. The entire, step-by-step procedure is illustrated in Fig. 15-4.

Winds aloft can be obtained by running a wind problem backwards. To solve for winds aloft you need: True Heading, TAS, wind correction angle, and ground speed. This information is available from cockpit indications (TAS, heading) and from the radar procedure just described (track, ground speed).

These techniques do require proficiency with the E6-B, and they may even require that a written description of the procedure be carried in the aircraft. (As a practical matter, very few pilots are going to remember how to calculate track and winds aloft using radar fixes unless they practice it regularly, and it is unrealistic to think that many pilots will do that

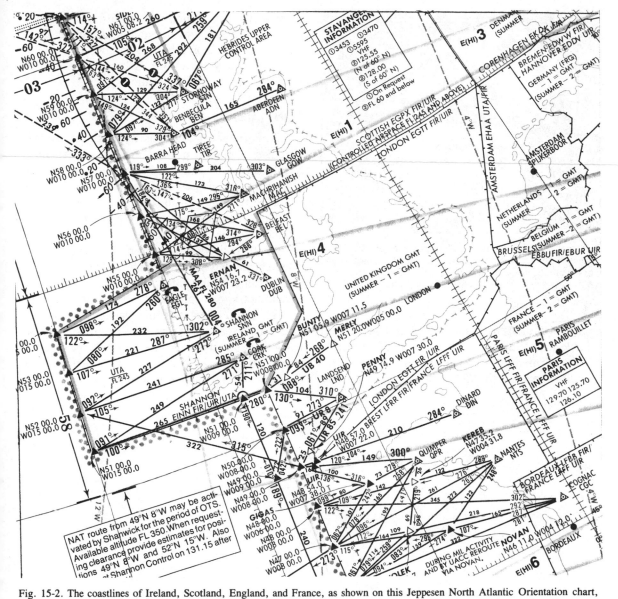

Fig. 15-2. The coastlines of Ireland, Scotland, England, and France, as shown on this Jeppesen North Atlantic Orientation chart, are distinctive and can be easily identified on the radar screen. Reproduced with permission of Jeppesen Sanderson, Inc. Not actual size. Not to be used for navigation.

given the long odds of its ever being needed.) The procedure is no different than that for obtaining ground speed and winds aloft data based on VOR/DME fixes and courses, however, and both procedures are valuable additions to an instrument pilot's bag of tricks—airway fixes may be the only way to obtain updated ground speed and winds aloft

data over land, and radar observations may be the only way over water.

If you haven't worked a winds aloft en route problem in awhile, this might be a good time to try one. The procedure is not at all difficult, but it does take practice to stay proficient. Winds aloft are a significant in-flight parameter and even if you never cal-

Fig. 15-3. Azimuth markings at 30 degree intervals help in estimating relative azimuth off the nose of the aircraft. Azimuth markings are turned on and off with the switch at the bottom marked AZ. Courtesy of Collins Avionics, Rockwell International.

aloft, on the assumption that any aircraft with area nav capability can provide winds aloft data automatically, and you should be able to give ATC an answer, even if your answer is based on the E6-B system of long-range navigation.

CONCLUSION

Airborne radar can be used for primary navigation, and is used for primary navigation in the military for low-level, all-weather operations. Military radar could be adapted to civilian use, but the gains would be relatively small, the risks would be high, and the increase in expense and weight would be substantial.

Civilian, airborne weather radar can be used for supplemental navigation though, and while it is important not to exaggerate the benefits, it is also important not to overlook them. Radar can be a very helpful tool in orientation and planning; it can help in the discovery and prevention of gross errors in navigation, and it can, in an emergency, be used for terrain mapping and dead reckoning when no other navigational information is available.

culate winds aloft from radar fixes you should at least be able to calculate them from VOR/DME tracks. ATC will often ask the pilots of aircraft on instrument flight plans with "/R" suffixes for actual winds

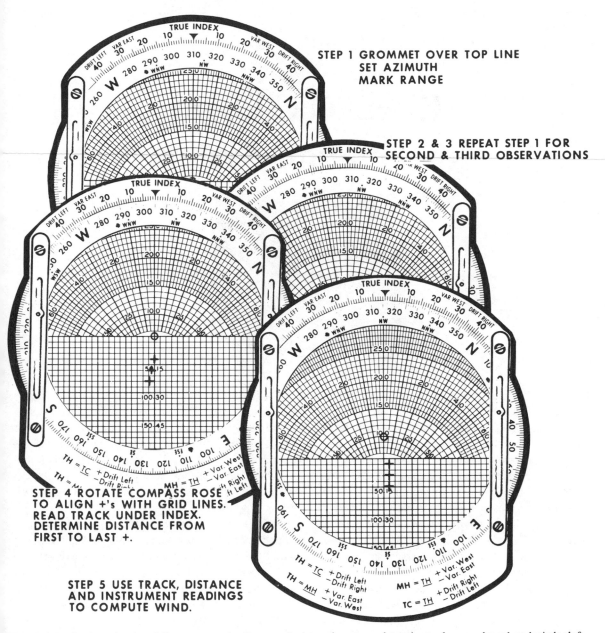

STEP 1 GROMMET OVER TOP LINE
SET AZIMUTH
MARK RANGE

STEP 2 & 3 REPEAT STEP 1 FOR
SECOND & THIRD OBSERVATIONS

STEP 4 ROTATE COMPASS ROSE
TO ALIGN +'s WITH GRID LINES.
READ TRACK UNDER INDEX.
DETERMINE DISTANCE FROM
FIRST TO LAST +.

STEP 5 USE TRACK, DISTANCE
AND INSTRUMENT READINGS
TO COMPUTE WIND.

Fig. 15-4. By plotting azimuth and distance to a radar fix at regular intervals, one can determine track, groundspeed, and winds aloft.

Chapter 16

Electronic Flight Information Systems

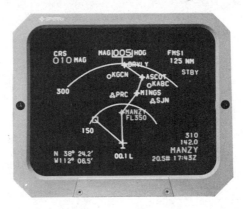

Electronic Flight Information Systems, or EFIS systems, are the leading edge in aviation technology for the presentation of aircraft attitude, navigation, hazardous weather, and flight direction. First introduced at the airline level on the Boeing 757 and 767, and at the corporate jet level on the Canadair Challenger and Falcon 50, EFIS systems are now available for a variety of turbine-powered aircraft, including the Beech King Air, the Mitsubishi MU-2, and the Piper Cheyenne series of turboprops. EFIS systems should be available at the piston twin and heavy single-engine level in the not too distant future. There is, in fact, every reason to believe that EFIS systems will eventually be available at all levels of aviation, including training and sport aviation: the benefits of electronic displays, when combined with the economics of volume production, should eventually result in even basic attitude and heading indicators being converted to electronic formats.

As with most new systems, the first electronic displays were greeted with considerable skepticism and wariness in the aviation community. There were, in fact, good reasons for this initial skepticism: most

new systems do have flaws and unforeseen drawbacks, and most pilots are reluctant to give up a system they know for one they do not know. In the case of Electronic Flight Information Displays (which sounds like "television" to most pilots), this initial wariness was further strengthened by direct experience with the vulnerability of TV tubes to damage and failure.

While these fears were at least partially justified initially, they have proven to be unfounded in the long run: the manufacturers and the FAA have developed safeguards and back-ups both for tube and for computer failures, and the results have been impressive. Even the most skeptical of pilots have become believers after using EFIS systems for even a short period of time, and failure modes (called *reversionary modes*) have proven to be rarely needed, but perfectly adequate when they were.

EFIS systems, while advantageous, are hardly necessary for safe navigation, and most operators wait for re-outfitting, overhaul, or new aircraft purchase before installing or purchasing an EFIS system; nonetheless, while the transition from mechanical to electronic systems may be slow, the trend is clear, and many pilots want to learn what they can about EFIS systems now, even if they do not anticipate actually having access to an EFIS system for some time. The purpose of this chapter is to provide that introduction—to give pilots a good idea of what EFIS systems are, what they can do, and why they are such valuable additions to the IFR cockpit in particular.

GENERAL DESCRIPTION

In the most basic sense, EFIS systems simply replace the conventional, mechanical Flight Director (FD) and Horizontal Situation Indicator (HSI) in a fully equipped IFR aircraft with computer generated, color digital, electronic displays. In itself this is advantageous, since the electronic display (a Cathode Ray Tube, or CRT) has no moving parts, and this greatly increases its reliability over comparable mechanical displays. (Normally, the existing gyros are retained in an EFIS installation, which leaves some moving parts in the total flight/navigation system; however, if a ring laser gyro has been installed—the state-of-the-art in gyro technology—then even that component is eliminated as a moving part, creating a totally solid state flight control and navigation system.)

EFIS systems do much more than simply replace mechanical indicators with electronic displays though. The EFIS system takes raw flight and navigation data—attitude, heading, VOR (or any other nav system) course and track, weather radar information, plus a variety of supplemental data such as DME distance, airspeed, and radar altitude—and directs that information to a specialized computer called a "Symbol Generator." The Symbol Generator looks at the type of display requested by the pilot—full compass, segmented compass, en route configuration, approach configuration, with or without weather radar overlay—sorts, categorizes, and analyses all the available flight and navigational information, and then converts that information into digital data from which color symbols similar to conventional FD and HSI symbols are created for display on the corresponding CRT.

This process of collection, analysis, and symbol generation is repeated, typically, 60 to 80 times per second. This is a much faster renewal rate than that necessary simply to respond to changing conditions, but it is made necessary in order to avoid a phenomenon known as flicker effect: it has been determined that a picture that is updated less than about 40 times per second will appear to flicker as it changes (like the old-time movies, and for the same reason); however, at rates greater than 40 times per second, the eye (more accurately, the brain) is unable to see the screen change, and the resulting display will appear to be continuous. The complexity of the computer programming necessary to handle this volume of information at these rates is mind boggling, but the result to the pilot is the same as if he were looking at a mechanical FD and HSI, however, the electronic display is steadier, clearer, and provides more information.

Nearly all pilots who have made the transition from mechanical to electronic displays have said that the process is perfectly natural and logical, partially because the Symbol Generator is invisible to the pilot, automatically working its magic in ARINC boxes hidden from view and not requiring any pilot manipulation, and partially because the symbols generated are nearly identical to what pilots are used to seeing on their mechanical FD and HSI displays. Few who have used EFIS systems ever want to go back to mechanical displays with their diminished reliability, reduced information capacity, lack of flexibility and redundancy, and relatively sluggish and coarse responses.

The typical EFIS system breaks down into two main components, each of which corresponds to the two main cockpit display components common to any Flight Director System: the Electronic Attitude Director Indicator (EADI) and the Electronic Horizontal Situation Indicator (EHSI). A third, optional display found in many EFIS systems is the Multifunction Display (MFD), a tube reserved for the display of additional, supplemental information—information that would be inappropriate for display directly on the EADI or EHSI itself.

A full system, with dual EADIs and dual EHSIs installed, one set per side, plus Multifunction Display, represents a total of five CRTs, or tubes, and a full system is therefore commonly referred to as a "five-tube system." (The MFD is often called the Fifth Tube.) An all-glass cockpit normally has yet another tube for the display of engine and aircraft systems instrumentation—a six-tube system. Manufacturers are, however, beginning to experiment with larger tubes that combine functions—a single tube for both the EADI and the EHSI for instance—and the practice of describing systems by the number of tubes may no longer be meaningful.

The next three sections look at each of the major EFIS cockpit components—the EADI, the EHSI, and the MFD—in a little more detail.

THE ELECTRONIC ATTITUDE DIRECTOR INDICATOR

The Electronic Attitude Director Indicator corresponds to the normal Attitude Director Indicator (ADI) in a traditional Flight Director system, which in turn corresponds to the attitude indicator, or artificial horizon, on a basic flight instrument panel. A typical EADI is illustrated in Fig. 16-1.

In most respects, an EADI looks like an ADI: blue background for sky, brown for earth, attitude or pitch markings above and below the horizon, a wedge shaped aircraft symbol, and pointers and scales for glide slope, angle-of-attack, etc. The similarity is, of course, intentional, in order to facilitate the pilot

Fig. 16-1. The Electronic Attitude Director Indicator (EADI) corresponds to the ADI in a Flight Director system. Copyright Sperry Corporation, 1985.

transition from mechanical to electronic displays, and to minimize any possibility for misinterpretation of this vital information.

The main advantage to the EADI over the conventional ADI is that the EADI can be reconfigured, at the pilot's command and as the situation dictates, whereas the ADI display (with very few exceptions), is fixed at the factory. (The glide slope needle portion of a conventional ADI, for instance, can be covered by a flag when not required, but it cannot be switched from one side to the other, and the space it takes cannot be used for other data when glide slope information is not required.)

In the example illustrated, the EADI has been configured for the approach mode, as indicated by the LOC and GS abbreviations at the top of the display; in this configuration, information unique to the approach environment is displayed on the EADI, and unnecessary or irrelevant information (such as Mach number or DME distance when DME is not available) is deleted. In the approach configuration, for instance, selected Decision Height is displayed in the lower left-hand corner of the EADI; in the cruise configuration, Decision Height would be deleted. Actual radar altitude (absolute altitude above the terrain as reported by the radar altimeter) is annunciated at the far right (140 feet AGL in this example) whenever the aircraft absolute altitude falls within radar altimeter limits (normally 0 to 2,500 feet AGL), and a lighted *DH* symbol is annunciated when that altitude has been reached. An electronic runway symbol is generated to provide orientation and alignment with the runway and an expanded localizer deviation indicator (essential for CAT II and III operations) replaces the electronic rate-of-turn indicator at the bottom center position whenever a Localizer frequency has been tuned. Passage over the outer, middle, and inner marker beacons is annunciated on the EADI by letters corresponding to the beacon identification, supplementing the traditional blue, amber, and white lights (located elsewhere on the panel). Digital indicated airspeed and trend information is displayed on the the left hand side of the primary display, and glide slope deviation is indicated by a pointer and scale on the right hand side. At cruise, most of these additional bits of information unique to the approach environment would be deleted.

The ability to optimize the display configuration for the particular flight condition—maximizing the amount of information on the approach, and eliminate unnecessary information en route—is one of the main advantages of the EFIS system.

THE ELECTRONIC HORIZONTAL SITUATION INDICATOR

The versatility that is the key characteristic of the electronic format is even more apparent with the EHSI than it is with the EADI. Conventional HSIs have always promised a little more than they delivered: conventional HSIs are limited to a description of aircraft heading and aircraft orientation with respect to a single selected course (normally a VOR radial). To be totally accurate, they should be called Partial Horizontal Situation Indicators. In addition, HSIs are limited to a single type of display, and the ability to reconfigure the display for the particular flight condition is limited in the same way it is for the mechanical ADI. (HSIs are still very useful devices; the EHSI is simply more so.)

The EHSI does not suffer from these limitations, but can be configured, at the pilot's command, between two different types of displays, and extended route information, as well as weather radar information, can be superimposed on the EHSI screen.

Full Compass Configuration

The EHSI in Fig. 16-2, for instance, is configured to the normal, full compass HSI presentation. In this configuration, up to two RMI pointers (a diamond-arrow pointer and a small circle-line pointer), corresponding to any combination of VOR or ADF bearing information, can be displayed directly on the EHSI screen. (Both can also be deleted entirely if not needed, simplifying the display.) In this example, the circle-line type pointer has been assigned to the Number 1 VOR and the diamond-arrow type pointer has been assigned to the ADF; the pilot is reminded of the RMI selections by the legend in the lower left hand portion of the screen. RMI pointer selection is done with a "Source Controller," illustrated in Fig. 16-3. In a dual installation, each side

Fig. 16-2. In the full compass configuration, the EHSI corresponds to a conventional HSI. Copyright Sperry Corporation, 1985.

normally has its own Source Controller. In this way, each pilot can assign course and bearing information independently for display on his EHSI.

In addition to the digital course and ground speed readouts available on most HSIs, the EHSI can also display certain background bits of information, including heading source (MAG1 at the top left on the EHSI illustrated indicates that the source for heading information is the Number 1 Magnetic Compass System), selected heading (HDG 315—this is command, or bug heading—annunciated at the bottom left), and nav source (NAV 1 means the primary EHSI course deviation needle is being driven by the Number 1 VOR—seen at top right). Selected course is digitally displayed at the top left, with weather ra-

Fig. 16-3. The Source Controller assigns navigational information to each of the RMI bearing indicators, as well as to the primary EHSI Course Deviation Indicator. Copyright Sperry Corporation, 1985.

dar antenna tilt angle (two degrees down) underneath that, and DME distance and ground speed on the right-hand side (top and bottom respectively). Supplemental, nice-to-know information such as this is easy to display around the margins of an EHSI, but would be impractical on a mechanical HSI—the conventional HSI lacks sufficient interior space to handle all the drums and gears necessary to display this information mechanically.

Segmented Arc Configuration

While all of this is useful, if the only advantage to an EHSI were the ability to display extra information around the margins of the display and provide a dual RMI pointer overlay, there would be little reason to justify its development, and in fact the EHSI is capable of much more. With the push of a single button the EHSI can be reconfigured from the traditional HSI type of display just described to a segmented arc display of 90 or 120 degrees (depending on the manufacturer), over which complete route and weather information for the range selected can be superimposed (see Fig. 16-4).

In the segmented arc configuration aircraft position relative to the selected route ahead and relative to any hazardous weather can be displayed with

great clarity and without sacrificing any essential nav data: course deviation information is still available in the form of an abbreviated Course Deviation Indicator, and actual aircraft heading can be seen at the top of the segmented arc. Any number of other bits of information, such as selected course (CRS 315), selected heading (HDG 319), digital readout of aircraft heading (321), ground speed (GSPD 260 KTS), and DME distance (130 NM) can also be displayed around the periphery, as appropriate to the particular situation.

In the segmented arc configuration, the EHSI becomes, in effect, a moving map display, providing a continuously updated picture of aircraft position relative both to the desired course and to any hazardous weather. In this configuration, display range is determined by the weather radar and is therefore limited to the maximum radar range (normally 200 to 300 nautical miles).

Display Control Panel

The ability to re-configure the EHSI gives the pilot the best of both worlds: the conventional HSI display for airway and approach navigation, the segmented arc display for long-range navigation and overlay of weather information relative to the route ahead. A Display Controller (Fig. 16-5) is used to switch between the two types of displays as desired (specifically, with the button labeled FULL/ARC). The round knobs to the left and right are used to slew, or change, course and heading pointers. Each side of a dual EFIS installation normally has its own Display Controller, allowing each pilot independent control over his display configuration, as well as allowing

Fig. 16-4. In the segmented arc configuration, weather radar information can be superimposed on the EHSI display, clearly indicating aircraft position relative to the route of flight and to any hazardous weather. Copyright Sperry Corporation, 1985.

Fig. 16-5. Control over all EFIS display formats, as well as control over course and heading pointers, is accomplished with the Display Controller. Copyright Sperry Corporation, 1985.

the Pilot-in-Command the option of configuring one side of the cockpit one way and the opposite side in another.

THE MULTIFUNCTION DISPLAY

The Multifunction Display (MFD) is an additional tube for the display of a variety of supplemental information. (See Fig. 16-6.) It is normally located in the middle of the cockpit where both pilots can see it. Selection of MFD functions, cursor control, range selection, and page selection are all controlled with a separate MFD Controller, normally located directly under the MFD display (illustrated in Fig. 16-7).

The main functions of the MFD are: mapping, planning, weather radar reporting, checklist readout, and CRT/Symbol Generator back-up. Each of these functions is worth an individual look.

Mapping Mode

The MFD displays much the same information in the mapping mode as the EHSI does in the segmented arc configuration, including overlay of weather information; however, the MFD is capable of displaying extended route systems with full navaid identifiers and lat/long coordinates for selected way-

Fig. 16-7. The MFD Controller is used to select MFD functions and to manipulate information for display. Copyright Sperry Corporation, 1985.

points, as well as supplemental information from other, optional systems such as from Flight Management Systems (see the next chapter) and from navigation data bases (a computerized Jeppesen library). The mapping mode is illustrated in Fig. 16-6.

Whenever the mapping mode has been selected, the MFD display will be oriented with the aircraft heading at the top and the aircraft position symbol at the bottom—the plan moves under the aircraft symbol as the flight progresses, just as it does on the EHSI segmented arc display. Range on the MFD can be extended well beyond that of the EHSI however (which is limited to whatever range the radar has been set to)—up to 1,200 nautical miles in the case of the Sperry system illustrated—allowing all but the very longest routes to be displayed in their entirety.

In addition, by positioning a moveable cursor anywhere on the screen, the MFD will automatically compute the lat/long coordinates for that position, and these coordinates can then be directly entered as the next waypoint. Cursor movement is controlled by the joystick at the top center of the MFD Controller, Fig. 16-7.

Aside from the convenience of this arrangement in designating checkpoints, this also greatly simplifies navigation around weather: with weather radar displayed on the MFD, the pilot can position the cursor between cells and then load the coordinates indicated into the computer as the next waypoint. With an appropriate clearance to those coordinates, the pilot can then proceed exactly to that point, rather than having to guess at the number of degrees to the right or left to deviate around the weather.

If a navigation data base has been installed, then direct courses to any point can be entered in this way

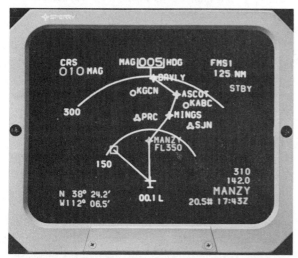

Fig. 16-6. The Multifunction Display can display additional navigational information in a variety of different formats, as well as provide back-up for the rest of the EFIS system in the event of CRT or Symbol Generator failure. Copyright Sperry Corporation, 1985.

as well, including direct courses to outer markers, airports, initial approach fixes, and intersections; there is no need to look up frequencies, identifiers, or coordinates, since these are all stored in the MFD computer and all fixes that fall within the range selected are automatically plotted on the display according to their lat/long coordinates.

In general, the MFD works best when it supplements the EHSI by displaying the big, en route picture, leaving the EHSI free to display airways directly ahead and approach courses. During an instrument approach for instance, the pilot would normally configure the EHSI to the conventional, full compass HSI type display, and configure the MFD in the mapping mode. In this way the MFD would serve both as a back-up to the approach display, and as an overview of the aircraft position relative to the entire approach environment: the Final Approach Fix, the airport, the Missed Approach Fix, and the route to the landing alternate. The approach itself would be conducted using the EHSI.

Planning Mode

In the planning mode, the MFD is oriented with north at the top (instead of aircraft heading); the aircraft symbol is positioned in the middle of a 360-degree plan view of the programmed route, pointed in the direction of flight, indicating present position and orientation with respect to the flight planned route. (See Fig. 16-8.) In this type of display—a fixed map display—the aircraft symbol travels along the route depicted as its position is updated.

The planning mode is best for general flight planning, for aircraft orientation relative to the flight planned route, and for an overview of the entire navigational scenario. The planning mode provides yet another way of looking at the aircraft position relative both to the cleared route and to alternate routes, and increases the versatility of the system.

Weather Radar

In both the all-glass and in the five-tube cockpit, the MFD also serves as the weather radar display. All of the electronics for the weather

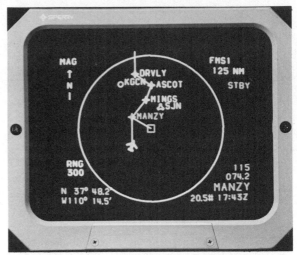

Fig. 16-8. The MFD planning mode is used for flight planning and for general course orientation. In this mode, north is always at the top; the aircraft symbol moves across the plan as the aircraft position changes. Copyright Sperry Corporation, 1985.

radar—transmitter/receiver, wave guide, digital computer—are remotely located, and the picture is fed not to a radar screen, but to the MFD. The weather radar becomes another source of information.

It is sometimes hard for pilots to understand how the radar can become simply another source of information, but this is in fact what it is: weather depiction is part of the information that is often necessary for the safe conduct of the flight, and in most cases the best place for that information is superimposed on the EHSI and MFD displays. Superimposing weather data on the nav displays is the only good way to visualize aircraft position and planned route relative to the weather depicted.

The MFD can also be used to study the weather itself by de-selecting both the map and the plan overlays, in which case it acts exactly like a conventional radar display (only with better clarity and definition than a conventional radar). The pilot might want to de-select the map and plan overlays in order to study a complex weather system in detail, for instance. Since, with an MFD, there is no actual radar set, a separate panel is provided for the control of all radar functions: range selection, gain control, mode selection, tilt and so on. An example is illustrated in Fig. 16-9.

Fig. 16-9. The Weather Controller provides control over all weather radar functions for display on either the EHSI or on the Multifunction Display. It corresponds to the controls on a conventional weather radar Control-Display Unit. Copyright Sperry Corporation, 1985.

Checklist Mode

All normal, abnormal, and emergency checklists can be stored in the MFD computer for display on the MFD CRT. (See Fig. 16-10.) Checklists can be manufacturer-supplied to the specifications of the particular aircraft type, or customized by the pilot or Chief Pilot to conform to individual aircraft and specific company policies and procedures.

The MFD checklist is more than convenient—it also helps guard against errors in checklist completion: completed checklist items are noted by a change in color; skipped items are flagged for later recall; the active item is enclosed in a box. In an emergency

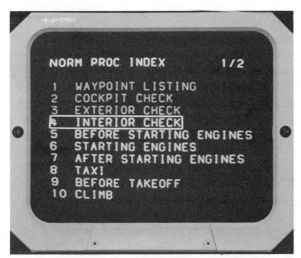

Fig. 16-10. The MFD computer is capable of storing all normal, abnormal, and emergency checklists for display on the MFD screen. The active item is enclosed in a box. Completed items are indicated by a change in color; individual items can be flagged for later recall. Copyright Sperry Corporation, 1985.

or abnormal situation, appropriate checklists can be called up in a matter of a very few seconds (versus the many seconds sometimes required to find a printed checklist buried in a pocket or hiding under the seats). With up to 400 pages of storage capacity, there is almost no limit to the amount of detail that can be included on these checklists, including the reproduction of schematic diagrams for individual aircraft systems.

EADI/EHSI Back-up

The MFD CRT can also serve as a back-up tube in the event of an EADI or EHSI tube failure, and the MFD Symbol Generator can be used as a spare Symbol Generator for the primary EFIS system or systems. The MFD thus greatly increases the redundancy and reversionary capabilities of any EFIS system, and is a very worthwhile addition for this reason alone. This aspect of MFD capability is explored in more detail in the next section.

REVERSIONARY MODES

Reversionary, or back-up, modes are an integral part of all EFIS system designs. Their purpose is to prevent the loss of key information in the event of a tube or Symbol Generator failure. The degree to which the display is compromised after a failure depends on the type of failure and the degree of redundancy in the total EFIS system: a dual installation has more redundancy, and greater reversionary capability, than a single installation, and a dual installation with Multifunction Display has greater reversionary capability yet.

Tube Failure

In a dual sided system, the failure of an EADI or EHSI tube on the side of the pilot not flying can usually be ignored. The situation is a little more serious when a tube fails on the side of the pilot flying, and it is quite a bit more serious yet in a single sided system, especially if it is the EADI that has failed. Reversionary modes to compensate for tube failures are, however, built into all systems, regardless of the number of tubes or systems installed.

In the event of an EADI or EHSI tube failure,

Fig. 16-11. In the event of EADI or EHSI tube failure, information from the failed tube is automatically directed to the remaining tube to form a composite picture that emphasizes aircraft control. Copyright Sperry Corporation, 1985.

most systems automatically revert to a Composite Mode in which the information from the failed tube is combined with information from the functioning tube for display as a composite picture of essential information. An example of a composite EADI/EHSI display is shown in Fig. 16-11. Note that attitude and Flight Director information is left essentially intact in the Composite Mode, while navigation information is reduced to the minimum necessary for safe flight: an abbreviated CDI along the very bottom and a rotating drum type of heading indication with digital readout. In the Composite Mode, priority is always given to aircraft control—attitude and Flight Director information. Basic navigational capability remains available, but in abbreviated form.

Symbol Generator Failure

Normally, a single Symbol Generator drives both the EADI and EHSI; therefore, if the Symbol Generator fails in a single sided installation, both parts of the EFIS system—EADI and EHSI—will be lost, and the pilot will be forced to revert to some combination of copilot's analog (mechanical) instrumentation, emergency standby gyro, VOR CDI, needle-ball-and-airspeed, or magnetic compass, as necessary, for air-

craft control and navigation.

In a dual installation, however, a separate Symbol Generator normally drives each side of the dual EFIS system, and either Symbol Generator can drive both sides if necessary. Thus, if a single Symbol Generator fails, the remaining Symbol Generator can supply both sides, and both pilots will continue to have fully functioning EADIs and EHSIs. The ability to compare systems will have been lost after a Symbol Generator failure, as well as the ability to configure one side in one way and the other side in another, but these are relatively minor considerations that affect the flight only until the next landing (when the failed Symbol Generator can presumably be removed by maintenance personnel and replaced with a functioning unit).

For all practical purposes, a dual EFIS installation allows both sides to continue to function after a single Symbol Generator failure without any compromise in the amount of information or in the clarity of the display. In the unlikely event of a dual Symbol Generator failure, reversion to emergency instrumentation—emergency horizon, mechanical RMI, magnetic compass—will be necessary. (It is obviously impossible to insure against never having to revert to emergency modes.)

MFD Reversionary Modes

The MFD increases reversionary capability substantially, since the MFD can function both as a spare tube and as a spare Symbol Generator. By electronically linking all systems together, the MFD serves as a general back-up to each of the other individual EFIS components.

The MFD CRT, for instance, can substitute, directly or indirectly as the case may be, for a malfunctioning EADI or EHSI display: it substitutes directly for the EHSI in the event of an EHSI tube failure, and it substitutes indirectly for an EADI failure by taking on the functions of the EHSI, freeing that tube to substitute for the EADI. This is a little confusing, but there is a good reason for it: this arrangement ensures that primary flight control information—the EADI display—will always be in front of the pilot flying, even if his EADI fails. (When the EADI fails, the EHSI on that same side becomes the EADI, and the MFD takes over for the displaced EHSI.)

In addition, the MFD Symbol Generator can be used as a spare to supply either or both sides of a dual EFIS system, and to back-up the primary Symbol Generator in a single sided system. In fact, in order to ensure that there is no interruption in the presentation of vital flight information, the MFD computer continuously monitors the other Symbol Generators, automatically taking over any time it detects an abnormality in either of the primary generators. In the language of the computer engineer, the failure of a primary Symbol Generator—the heart of the system—is transparent to the pilot: the MFD Symbol Generator is always on standby and automatically takes over without the pilot having to do anything. The pilot normally will not even be aware of the failure, except through an annunciation of Symbol Generator failure on one of the screens.

THE LONG TERM PROSPECTS FOR GENERAL AVIATION

EFIS systems are already available for nearly all turbine aircraft in sizes, configurations, and costs that are appropriate to the aircraft type. EFIS systems are so much more capable, reliable, and flexible than comparable mechanical Flight Director and HSI systems, that it is a virtual certainty that the trend toward EFIS systems as standard equipment on new aircraft will continue, and will not be limited to turbine powered aircraft.

Ultimately, there is no reason why the basic EFIS format—a color CRT driven by a computer chip Symbol Generator—could not be adapted down to the level of even basic flight instruments: attitude indicators, directional gyros, and course deviation indicators. There is no practical reason, for instance, why the entire package—color CRT, simplified Symbol Generator, and vertical or horizontal gyro—cannot be assembled into an individual panel mounted unit similar in size, weight, and cost to existing attitude and directional gyros. It is very possible, in fact, that conventional attitude indicators and directional gyros will look as obsolete ten or fifteen years from now as the old black-and-white artificial horizons and counter-rotating, drum-type directional gyros do today.

However reluctant pilots may be to accept these computer driven video displays for flight instrumentation initially, those who have used them have become instant converts, and the trend is clear, unmistakable and unstoppable: the modern IFR aircraft cockpit is going to be an EFIS cockpit, at least in part, and the modern VFR cockpit will almost certainly include EFIS components. It is not too soon to begin anticipating these changes.

Chapter 17

Navigation and
Flight Management Systems

As long-range navigation systems multiplied it became clear to pilots that a consolidation of the control and display functions for all the various systems was in order—it wasn't necessary for each long-range system to have its own CDU, in fact it was actually somewhat cumbersome that it did. It also became clear that no single nav system was perfect all the time and that the key to accurate long-range navigation was to be able to take advantage of the strengths of each system as appropriate to the particular situation. What was less clear initially, but what also became clear in time, was that it was possible with proper nav system management to achieve an overall level of accuracy that was greater than the accuracy of the single best system at any given moment.

What was needed, actually, was a super-system: a system that consolidated and managed each of the individual systems into what appeared to be a single long-range system that was at least as accurate as the best individual system at any given moment. The first such systems appeared around 1982 and were called Navigation Management Systems (NMS). Later, first on the Boeing 757/767, and then on the Airbus

A-310, data from a flight performance computer—the digital equivalent of the charts and graphs in the aircraft flight manual—was integrated into the Navigation Management System, and the system that resulted from that combination was called a Flight Management System (FMS). Flight Management Systems combine navigational data with stored performance and aerodynamic data to determine a flight profile that is optimized for one of four operational parameters: speed, range, endurance, or constant Mach.

In simplest terms, Navigation Management Systems are navigators' navigators, and Flight Management Systems are pilots' pilots: an NMS helps the navigator/pilot find his way through a sometimes conflicting array of navigational data, and an FMS leads a pilot through a profile of his flight—takeoff, climbout, cruise climb, initial level-off, step climb, cruise, top-of-descent, descent, approach, and landing—a profile that has been optimized according to the parameter specified.

Neither Navigation Management Systems nor Flight Management Systems are simple systems, nor are they inexpensive, but, for those who can afford them (or for those who cannot afford not to have them), they work a certain magic, extracting, in some paradoxical way, more information than is possible from the sum of the individual systems. NMS and FMS systems could well be the most logical and useful systems to have been developed for internationally operated aircraft since the development of INS and OMEGA/VLF navigational systems themselves.

NAVIGATION MANAGEMENT
—THE PROBLEM

Navigation Management Systems are solutions, but before proceeding with the solution, it is important to first fully understand the problem.

The problem is actually very simple—there is no such thing as a perfect navigational system: NDB is very inexpensive, but its range is variable and unpredictable, and accuracy is minimal; VOR is easy to use and inexpensive, but its range is limited and it is not especially accurate; DME is quite accurate, but only at short ranges; OMEGA/VLF provides good long-range coverage, is relatively consistent, but it is only moderately accurate; LORAN-C is very accurate, at times, but its accuracy is somewhat inconsistent, and its coverage is limited; INS is extremely accurate short term, but it is extremely expensive and its accuracy deteriorates with time; NAVSTAR/GPS has much promise, but it is unproven, and it may be years until it is fully operational.

If there were a perfect navigational system, then the solution would be obvious (assuming cost were not a factor): install as many perfect systems as are required for full redundancy and dispatch reliability (probably three), and be done with it. (This is the reasoning behind triple INS installations; INS may not be a perfect system, but it is the closest thing to it at this time.) Since the perfect system does not exist, the next most logical approach (again, cost considerations notwithstanding), is to install at least one of each type of navigational system: VOR/DME, OMEGA/VLF, LORAN-C, INS, and NAVSTAR/GPS (when it becomes available). Since LORAN-C coverage is limited at the present time, operators desiring triple, worldwide nav redundancy

would have to have either an extra OMEGA/VLF or (preferably) an extra INS. Thus the "perfect" state-of-the-art cockpit would have: dual VOR/DMEs, a LORAN-C unit, an OMEGA/VLF unit, dual (or possibly even triple) INSs, and space and wiring for the addition of a NAVSTAR/GPS system when that becomes available. (NDB equipment would of course also be installed for approaches and back-up.)

With this many systems installed, simply finding space for all of the CDUs would be very difficult, and even if space could be found, the workload for the pilot/navigator would be overwhelming: he would have to continuously monitor all systems, knowing when to reject certain systems, when to rely on other systems, when it was best to average several position estimates, and when and how to update one system with another. The closer you get to creating an ideal nav system, the closer you get to reaching a practical limitation on cockpit space and pilot workload. That is the problem.

NAVIGATION MANAGEMENT
—THE SOLUTION

Navigation Management Systems attack the problem of space and pilot workload simultaneously. The space problem is the easiest to solve.

Solving the Space Problem

There is almost always room on an airplane for another system if it can be put in a box and hidden away in a radio rack somewhere. By eliminating all but the receiver/processor portion of each nav system, and then connecting all of these receiver/processors (nav sensors) to a single NMS system with a single CDU, the space problem can be solved. With only one CDU in the cockpit, any number of individual nav sensors can be connected to it without impinging on cockpit space. (Dual NMS systems are very often installed anyway, for the usual reasons, but even two NMSs take up quite a bit less space than multiple individual systems.) The NMS acts like the preamplifier or receiver in your stereo system, consolidating and controlling a variety of sources (tuners, tape decks, turntables, and so on) through a single unit.

Most NMS systems have *ports* (input channels) for at least five sensors. Thus a typical NMS system could handle the following array of individual nav sensors:

1. DME
2. LORAN-C
3. OMEGA/VLF
4. INS
5. SECOND OMEGA/VLF or INS.

While five ports is typical, some of the earlier NMS systems were designed to work only with DME and one additional sensor; some of the newer systems can accommodate seven and eight sensors, including NAVSTAR/GPS. Obviously, the more capability the better, but simpler NMS systems may be fine for those operators who do not envision ever installing more than a single long-range system (typically OMEGA/VLF).

The Workload Problem

With all nav sensors piped into one CDU, control over the various systems is easy, but deciding which one or ones to use at any given moment is another matter. LORAN-C accuracy, for instance, is a direct function of LOP geometry, but unless you want to spend a lot of time plotting hyperbolas on a large LORAN-C chart, it is very difficult to know exactly what level of accuracy you are getting with LORAN-C: it could be as accurate as 0.1 nautical miles, or as inaccurate as 2.5 nautical miles, depending on aircraft position. INS will always be more accurate than OMEGA/VLF initially, but INS accuracy degrades with time; it is very difficult to pinpoint the exact point where the OMEGA/VLF position becomes more accurate than the INS position, because OMEGA/VLF accuracy is itself not a constant. Unless you know which system is most accurate at any given moment, there is no way to know which system is best to use for primary position information.

If DME is available, deciding which system to use is easy: DME is always more accurate than any long-range system, (consistently accurate to within 0.1 nautical miles), but only if it has been corrected for slant range error, and that is a laborious process to do manually. Even then, the theoretical accuracy is nearly impossible to achieve in practice, since at least three DME arcs have to be plotted on a chart to determine a fix, and even if those arcs coincide perfectly (which is unlikely), the width of the pencil line itself may be as much as a mile across, depending on the chart scale. Finally, once you do have a fix, converting it to a lat/long coordinate involves estimating and interpolating from adjacent scales, which introduces an additional error.

The workload problem is, in other words, more than just a problem with the amount of information that must be analyzed and processed, it is also a problem with the difficulty of manually analyzing and processing information to anything like its theoretical potential. Fortunately, the computer portion of the Navigation Management System is fast, smart, and careful, and can do difficult computations involving large amounts of information with great accuracy. The solution to the workload problem is therefore another computer.

The Important Role of DME

Since DME is the most accurate navigation system presently in use (with the very minor exception of INS for the first few minutes after takeoff), all NMS systems use DME first for position determination, regardless of whatever other information might also be available.

In an NMS system, the DME sensor automatically and continuously scans for DME facilities, and tunes four or more suitable stations—the exact number varies from manufacturer to manufacturer. It directs the resulting distance measurements to the NMS computer where they are corrected for slant range, internally plotted, and then converted into lat/long coordinates. Because of the number of calculations that must be performed repeatedly and rapidly, the only practical way to navigate using multiple DME fixing is to have a computer to do the work.

At this point, we have an NMS that serves as a control console for different nav systems with a built-in DME based (rho/rho) RNAV computer. To take the next logical step and extend the NMS computer beyond these two functions inevitably involves making design choices, and that is the point where easy solutions are left behind.

NAVIGATION MANAGEMENT METHODS

There are at least three different ways to go about processing navigational information from different long-range sources in order to arrive at a single most probable position: the Priority Based method, the Weighted Average method, and the Kalman Filter method. The methods vary in degree of sophistication and accuracy, but all accomplish the primary NMS objectives: a reduction in pilot workload and a greater degree of overall accuracy than that obtainable directly from individual systems.

Priority-Based Nav Management

The priority based method of nav management looks at all the available sources of nav information—DME, LORAN-C, OMEGA/VLF, and INS—and then selects the one system that seems most accurate to it at that moment. (VOR bearing information is generally ignored by all NMS systems, however a few systems are designed to utilize VOR information if no nav information is available. None use NDB.) It then computes a most probable position based on the system selected.

The most important part of this system is the ranking of priorities. Nav system priorities are not fixed (except for DME priority), but vary with the geographic location of the aircraft (which affects both signal strength and signal geometry), and with time (which affects INS drift). The greater the number of nav sensors, the more oblique the signal geometry, the weaker the signal strengths, and the longer the flight, the more difficult and critical is the task of establishing priorities.

DME is always the first priority: if sufficient DME information is available to compute a fix, the NMS system uses DME information and ignores the rest. (The other sensors remain active, but are not used unless required, i.e., unless DME signals are lost.)

If DME is not available, then the NMS looks at what is available and makes a decision as to the most likely best system at that moment; obviously, if only one other system exists, such as an OMEGA/VLF system, then that system is used. If more than one exists, then it goes through its sorting process, looking at signal strengths, LOP geometry, inherent accuracy, and probable drift rates, and picks the system most likely to have the greatest accuracy at that exact moment. It then computes a position fix based on the selected system. The Collins FMS-90 illustrated in Fig. 17-1 is a priority-based Navigation Management System that accepts inputs from VOR/DME sources and from a single OMEGA/VLF.

Even with a simple DME-OMEGA/VLF-based NMS system, overall accuracy is improved over individual systems since the OMEGA/VLF system starts with a very precise initial fix from the moment it takes over for the DME sensor. Most NMS systems also include programming to smooth the transition from long-range navigation back to short range

Fig. 17-1. The Collins FMS-90 Navigation Management System uses DME rho/rho based position fixing to supplement its internal OMEGA/VLF sensor. Courtesy of Collins Avionics, Rockwell International.

navigation, so that the Most Probable Position does not suddenly jump when DME is re-acquired after a long period of OMEGA/VLF or INS navigation.

Weighted Averaging

Increasingly, more and more priority-based systems are assigning a quality value to each source after sorting, and are then averaging, or blending the individual fixes based on their weighted value; in effect, the primary, first priority fix is fine-tuned by the lower priority fixes. This is a relatively simple form of integration that improves the accuracy of the best source fix. The Global NMS illustrated in Fig. 17-2 is an example of a system that uses blending to compute an averaged position.

Kalman Filtering

A more sophisticated method of nav management is based on an advanced form of statistical analysis called *Kalman Filtering*. (Kalman Filtering is to averaging as calculus is to analytic geometry.) Kalman

Fig. 17-2. The Global GNS-1000 Navigation Management System uses the information from up to four separate long range nav systems to arrive at a composite Most Probable Position. Courtesy of Global Systems.

Filtering is a mathematical method of combining the information from a variety of sources to arrive at a single best answer. The algorithms for a sophisticated Kalman Filter can go on for pages and can take four and five seconds for even the fastest computer to process. Kalman Filtering is the kind of thing that makes mathematicians' and electronic engineers' eyes light up, and others glaze over. It is, however, a highly effective method of blending similar but somewhat conflicting data into a single value.

Even sophisticated nav management systems based on Kalman Filtering techniques start with DME for position fixing—when DME is available, it will always be more accurate than any result based on Kalman Filtering. When DME is not available, the nav computer then looks at all the information that is available, directs that data to the Kalman Filtering circuits, and computes a position fix based on a composite of data from the available nav sources.

The specific techniques for manipulating that data vary from manufacturer to manufacturer—Kalman Filtering is a method, not a formula you can look up somewhere. In general, however, NMS systems based on Kalman Filtering use INS only for velocity measurements, while each nav system maintains its independence—all blending is done in the NMS system, and no one system is ever used to update another. In this way, individual system integrity is maintained, and navigational boot-strapping—one system telling another where it is, so that that system can tell the first system where it is—is eliminated. The Universal NMS illustrated in Fig. 17-3 is an example of a system that uses this particular form of Kalman Filtering to arrive at what Universal calls One Best Computed Position.

To summarize: All navigation management systems are generally capable of accuracies to within 0.1 nautical miles when DME is available. When DME is not available, some systems establish priorities for the use of supplemental nav systems, monitoring and selecting the optimum system for the particular situation in order to insure maximum available accuracy at all times. Others will blend, in ways that vary in sophistication and detail from manufacturer to manufacturer, all of the available nav information, in order to arrive at a single-position fix that will be

Fig. 17-3. The Universal UNS-1 Navigation Management System uses Kalman Filtering techniques to arrive at "One Best Computed Position." The larger ARINC box to the left houses the NMS computer. OMEGA/VLF and LORAN-C sensors are housed in the two smaller ARINC boxes on the right. Courtesy of Universal Navigation Corp.

more accurate than that produced by any individual system. These techniques take NMS beyond the realm of mere management of nav sources, and add yet another layer of accuracy to what would be available through nav management alone.

FLIGHT MANAGEMENT SYSTEMS

The phrases Navigation Management System and Flight Management System are often used interchangeably, and the distinction between Vertical Navigation and Flight Management is often a fuzzy one. In fact, there are real differences among these systems, and the distinctions are worth observing.

Navigation Management Systems are what we have been discussing so far: the management and integration of various nav sources into a single functioning unit for the purpose of reducing pilot workload and maximizing system accuracy. Both Vertical Navigation and Flight Management Systems

increase the capabilities of Navigation Management Systems in incremental steps.

Vertical Navigation

Vertical Navigation (VNAV) has been around, in a somewhat primitive state, for several years, but it had to wait for the development of Navigation Management Systems to realize its full potential. Vertical Navigation means the computation of a specific climb or descent rate in order to arrive at a predetermined altitude over a specific fix—"Cross XYZ VOR at 9000 feet" for instance—or in order to climb or descend at a given angle—a three-degree descent angle, for instance. (Vertical angle is not a simple function of pitch: wind affects climb and descent angles, as anyone who has ever encountered an unexpected tailwind while trying to clear an obstacle on takeoff knows.)

Older, dedicated VNAV computers never

achieved the popularity expected of them, partially because they contributed little relative to the expense, pilot training, and workload involved, and partially because they only worked well in one direction, and that was down—they should have been called descent-planning computers. (That is not to say that VNAV computers were not worthwhile as far as they went, just that few owners felt the additional expense was worth the small return, and few pilots bothered to learn how to use them or stay proficient in their use.)

With the introduction of navigation management systems, however, the situation changed a little. All navigation management systems receive altitude information in order to correct for DME slant range; with that information already available, little additional expense is involved in adding VNAV capability to an existing NMS system (an upgrading of the software is all that is necessary), and the additional pilot training and effort necessary to utilize the VNAV function is minimal—the cost/benefit ratio easily tilts to a positive value. Thus, many, if not most, existing NMS systems either include VNAV capability, or are in the process of adding it to their systems.

Flight Management

VNAV computers show the pilot how to achieve a given climb or descent angle and how much to increase or decrease pitch angle to arrive at a certain point at a programmed altitude, but they cannot tell the pilot whether that angle will result in the fastest climb rate, or the most efficient rate, or whether climb or descent is even advisable at that point or not. Vertical navigation knows nothing about the performance of the airplane. A flight management system does. (Some manufacturers, due to competitive pressures, call their navigation management systems flight management systems even though performance information is not included with the NMS. This is marketing hyperbole, and I suspect that these same marketing pressures will soon force them to discontinue this somewhat misleading practice.)

Permanently stored in the flight management system computer is a complete library of aircraft performance information. By combining performance data with navigational data—aircraft position, ground

speed, winds aloft, outside air temperature, fuel flow, TAS, and so on—the FMS can continuously recalculate the pitch and power settings necessary to achieve the programmed result entered by the pilot: minimum time en route, maximum range, maximum endurance, or constant Mach number. The pilot can get some of this information from the aircraft performance charts (long-range cruise charts, for instance), and flight planning computers can be used to minimize time en route or to maximize fuel efficiency, but only a flight management system that is installed on the aircraft with continuous access to changing information can achieve optimum results.

The FMS does all the complex number crunching necessary to determine, for instance, the best altitude for maximum speed or minimum fuel, the exact power setting for maximum range for any given aircraft weight and tailwind component (cruise control), the best power setting for maximum speed or minimum fuel for any assigned altitude, and projected step climb capability. It can calculate takeoff and landing numbers—field lengths, V speeds, flap settings, power settings, pitch attitude, even acceleration parameters—and, with an engine out, the FMS can tap the computer for single engine climb speeds, single engine ceilings, drift down altitudes, and power settings for maximum single engine speed, range, or endurance. Climb profiles can be obtained both for maximum rate and for several cruise climb schedules, and the FMS can determine optimum top-of-descent points given the calculated winds aloft at cruise and distance to go to touchdown. In its ultimate form, the information from the FMS can even be directed to an auto-throttles system (an autopilot capable of physically moving not only the flight controls, but also the throttles), resulting in even greater efficiencies: the auto-throttles system can set the power controls more accurately than a pilot can.

Flight Management Systems have to be separately developed for each aircraft type and powerplant for which they are certified, which accounts for a large part of their substantial cost—in physical terms, an FMS is just another computer, but the developmental man-hours are enormous. Flight management systems are currently available for the Boeing 757/767 and the Airbus A-310, and at the corporate jet level, true flight management systems are under development for the Falcon 900 and Gulfstream IV, with other large corporate aircraft types to follow. An example of an existing flight management system, manufactured by Sperry, is shown in Fig. 17-4. The navigation and performance computers are housed in separate one-half ATR sized ARINC boxes, and are remotely mounted.

No doubt, over the years, FMS systems will also be developed by other manufacturers and for other aircraft, down to and including piston twins and sophisticated singles (any aircraft regularly used for transportation is a potential candidate for an FMS), but it will be probably be some time before these systems are inexpensive enough to be practical on non-turbine aircraft. Nonetheless, while these systems seem to be incredibly exotic and expensive at this point, it is important to remember that at one time a bulky, analog DME with no ground speed capability was also considered to be an expensive, exotic

Fig. 17-4. The Control/Display Unit for the Sperry Flight Management System features separate alpha and numeric keyboards, as well as unique cursor selection keys along each side of the display CRT. Navigation and performance computers are housed in separate ARINC boxes. Copyright Sperry Corporation, 1985.

piece of nav equipment, one found only on the newest airliners: FMS systems for general aviation may arrive sooner than you think.

SATELLITE NAVIGATION AND NAV MANAGEMENT

There will almost certainly be a role for flight management systems for the foreseeable future, but one of the many uncertainties surrounding the eventual deployment and operation of the full NAVSTAR/GPS satellite system of navigation is the effect that system will have on navigation management systems.

NAVSTAR/GPS promises to provide continuous, predictable, worldwide accuracies to within 100 meters. By comparison, DME, the most accurate existing system, is only accurate to within 185 meters. Even so, DME is so much more accurate than any long-range system that all navigation management systems use DME for primary position fixing whenever it is available. If NAVSTAR/GPS is even more accurate than DME, and if NAVSTAR/GPS does not need to be supplemented by other systems over water, then, logically, NAVSTAR/GPS should be able to satisfy all navigational requirements in the future, and the need for separate nav systems, as well as the need for a navigation management system to handle those separate systems, will no longer exist. If, on the other hand, NAVSTAR/GPS does not work as well as expected, then navigation management systems will continue to be needed.

Manufacturers are presently preparing for both eventualities by designing existing systems to accommodate NAVSTAR/GPS sensors when they become available. If NAVSTAR/GPS works as well as expected, then it can be expected that the NAVSTAR/GPS sensor will end up doing nearly all the work—the other nav systems will go along for the ride, no doubt making themselves useful with reasonableness checks and standing by for back-up, particularly during periods of less-than-optimum satellite geometry, but otherwise not being required. If,

on the other hand, NAVSTAR/GPS does not work out as well as expected, then the NAVSTAR/GPS sensor will simply become an additional sensor, to be blended into the composite Most Probable Position according to ever more complex algorithms.

The future is always hard to see with any degree of clarity, and until we have some actual working experience with the NAVSTAR/GPS system it is impossible to tell whether Navigation Management Systems will be needed in the future or not. The most difficult decisions will have to be made by those responsible for specifying nav systems on new turbine aircraft in the next two or three years: they will have to decide whether or not they should invest an additional $100,000 plus (per side) for a system that may be unnecessary in the near future. The rest of us have the luxury of waiting to see.

FINAL CONCLUSION

With this peek into the future we conclude this guide to modern air navigation. While it is my hope that you have found the specific information in this book to be informative and practical, it is also my hope that you have gained an appreciation for some of the general principles common to all air navigation. This book ends with a short list of what, it seems to me, are some of the more important of these general principles:

1. Accuracy in air navigation has much more to do with the amount of knowledge, care, and planning the pilot brings to the operation than it does with the specific navigational systems installed.
2. There is no such thing as a perfect nav system, and there probably never will be. Every nav system has its strengths and its weaknesses; it is the pilot's job to capitalize on the strengths of each system, and to minimize the weaknesses.
3. A pilot cannot have too much navigational information.
4. Navigation is not an end in itself, however much it may appear to be when studied in detail—it is a means to an end. The end is safe flying.

Appendix A

Document Sources

NAUTICAL |₀ ₁₀ ₂₀ ₃₀ ₄₀ ₅₀ ₆₀ ₇₀ ₈₀|

The following documents are available through the U.S. Government Printing Office. Prices vary and must be requested in advance. Address all inquiries to:

> Superintendent of Documents
> U.S. Government Printing Office
> Washington, DC 20402

Airman's Information Manual, U.S. Department of Transportation, Federal Aviation Administration (Washington, published quarterly).

Instrument Flying Handbook, U.S. Department of Transportation, Federal Aviation Administration (Washington, 1980).

International Flight Information Manual, U.S. Department of Transportation, Federal Aviation Administration (Washington, published annually, quarterly supplements).

Air Navigation: Flying Training, AF MANUAL 51-40, NAVAIR 00-80V-49, Department of the Air Force, Department of the Navy (Washington, 1983).

The following booklets are published by the manufacturers listed, and are available by writing directly to them.

MLS Is Operational
W.C. Reed, Allied Bendix Aerospace (Baltimore, 1984):

> Bendix Communications Division
> 1300 East Joppa Road
> Baltimore, MD 21204

OMEGA Navigation Systems
Canadian Marconi Company (Montreal, 1985):

> Canadian Marconi Company
> Avionics Division
> 2442 Trenton Ave.
> Montreal, Canada H3P 1Y9

"An Introduction to GPS: Everyman's Guide to Satellite Navigation", Steven D. Thompson, ARINC Research Corporation (Annapolis, 1985):

ARINC Research Corporation
2551 Riva Road
Annapolis, MD 21401

CAROUSEL IV Inertial Navigation System: System Technical Description, Delco Electronics (Milwaukee, 1974):

Manager
Commercial Avionics Marketing and Sales
Delco Electronics
General Motors Corp.
Milwaukee, WI 53201

LORAN-C and II Morrow, II Morrow, Inc. (Salem, OR, not dated):

II Morrow, Inc.
P.O. Box 13549
Salem, OR 97309

"LORAN Report", ARNAV Systems, Inc. (Salem, OR 1984):

ARNAV Systems, Inc.
4740 Ridge Drive NE
P.O. Box 7078
Salem, OR 97303

UNS-1 Integrated Flight Management System: Technical Description, Universal Navigation Corporation (Torrance, CA 1985):

Universal Navigation Corp.
3545 W. Lomita Blvd. (Unit B)
Torrance, CA 90505

The following booklet is available directly from the FAA:

"Getting Ready for MLS", Federal Aviation Administration (Washington, 1985):

U.S. Department of Transportation
Federal Aviation Administration
Program Engineering and Maintenance Service
Washington, DC 20950

The following journal is the leading scholarly periodical for both marine and air navigation. There is a subscription fee:

Navigation, Institute of Navigation (Washington, published periodically):

Institute of Navigation
Suite 832
815 15th Street, NW
Washington, DC 20005

The following organizations can provide assistance with flight planning, weather briefing, and international operations:

Aircraft Owners and Pilots Association
421 Aviation Way
Frederick, MD 21701

Executive Air Fleet Corp.
Teterboro Airport
90 Moonachie Ave.
Teterboro, NJ 07608

Lockheed Dataplan, Inc.
90 Albright Way
Los Gatos, CA 95030

Universal Weather and Aviation, Inc.
8222 Travelair St.
Houston, TX 77061

Information concerning specific international navigational requirements can be obtained by writing:

The International Civil Aviation Organization
Place de l'Aviation International
P.O. Box 400
Montreal P.Q.
Canada H3A 2R2

Appendix B

Aeronautical Chart Sources

NAUTICAL |᠁᠁᠁᠁᠁᠁᠁᠁᠁᠁᠁᠁᠁᠁᠁᠁᠁᠁᠁|
0   10   20   30   40   50   60   70   80

Aeronautical charts for the U.S. and its territories and possessions are produced by the National Ocean Service (NOS), a part of the Department of Commerce, from information furnished by the FAA. Types of aeronautical charts available are:

Sectional and VFR Terminal Area Charts
World Aeronautical Charts (U.S.)
En Route Low and High Altitude Charts
Aircraft Position Charts
Planning Charts
Instrument Approach Procedure Charts
Standard Instrument Departure (SID) Charts
Standard Terminal Arrival (STAR) Charts
Airport/Facility Directory

A more complete description of charts that are available can be found in Chapter 8 of the *Airman's Information Manual*.

Information on obtaining NOS charts is available by writing to the:

National Ocean Service
NOAA Distribution Branch (N/CG33)
Riverdale, MD 20737

The leading commercial producer of aeronautical charts is Jeppesen Sanderson, Inc. Jeppesen chart subscriptions are normally sold through local dealers (Fixed Base Operators). Information can be obtained by writing:

Jeppesen Sanderson, Inc.
55 Inverness Drive East
Englewood, CO 80112
(303) 799-9090.

NOS charts are available directly from:

Sporty's Pilot Shop
Clermont Airport
Batavia, OH 45103
(800) 543-8633

NOS charts are also available from other mail-order suppliers. Most mail-order suppliers (including Sporty's) also carry a complete line of navigational aids, including plotters, computers, flight planning forms, navigation logs, clip boards, training aids, and instruction manuals.

Appendix C
Navigation Symbols

NAUTICAL

0   10   20   30   40   50   60   70   80

SYMBOLS — CHART AND NAVIGATION

————————	COURSE LINE
———————→	TRUE HEADING
– – –→ – – –	TRUE HEADING
———————→→	TRACK
———————→→→	WIND VECTOR
←————————→	LINE OF POSITION (LOP)
←←————————→→	ADVANCED OR RETARDED LOP
◄————————►	AVERAGE LOP
⊙	CHECKPOINT/NAVIGATION POINT
⊚	ALTERNATE/EMERGENCY AIRFIELD
▢	ORBIT POINT
▭	INFORMATION BOX
OAP symbol	OAP (OFFSET AIM POINT)
Course info symbol	COURSE INFORMATION BOX

+	AIR POSITION
⊙	DEAD RECKONING (DR) POSITION
⊙ MPP	MOST PROBABLE POSITION
▢	COMPUTER POSITION
△	FIX
△C	CELESTIAL
△L	LORAN
△M	MAP READING
△O	OMEGA
△R	RADIO
△V	RADAR
∧	CELESTIAL ASSUMED POSITION
↓ ∨ ⟩⟩	NO CHANGE FROM PREVIOUS LOG ENTRY

Appendix D

Abbreviations

AC	Advisory Circular.
ACT GS	Actual Ground speed.
ADF	Automatic Direction Finder.
ADI	Attitude Director Indicator.
AFR	Actual Fuel Remaining.
AGL	Above Ground Level.
AIM	*Airman's Information Manual.*
ANT	Antenna.
AOPA	Airplane Owners and Pilots Association.
ARINC	Aeronautical Radio, Incorporated.
ASF	Additional Secondary Factor.
ATA	Actual Time of Arrival.
ATC	Air Traffic Control.
ATE	Actual Time En Route.
BFO	Beat Frequency Oscillator.
BRG	Bearing.
C/A	Coarse/Acquisition.
CAS	Calibrated Airspeed.
CDI	Course Deviation Indicator.
CDU	Control-Display Unit.
CH	Compass Heading.
CLC	Course Line Computer.
CRS	Course.
CRT	Cathode Ray Tube.
CW	Continuous Wave.
DA	Drift Angle.
DEV	Deviation.
DG	Directional Gyro.
DH	Decision Height.
DIST	Distance.
DME	Distance Measuring Equipment.
DME/P	Precision Distance Measuring Equipment.
DOD	Department of Defense.
DR	Dead Reckoning.
DTK	Desired Track.
EADI	Electronic Attitude Direction Indicator.
EFIS	Electronic Flight Information System.
EFR	Estimated Fuel Remaining.
EHSI	Electronic Horizontal Situation Indicator.
ETA	Estimated Time of Arrival.
ETE	Estimated Time En Route.
ETP	Equal Time Point.

FAA	Federal Aviation Administration.	MNPS	Minimum Navigational Performance Specifications.
FAF	Final Approach Fix.	MOCA	Minimum Obstacle Clearance Altitude.
FAP	Final Approach Point.		
FAR	Federal Air Regulation.	MPP	Most Probable Position.
FD	Flight Director.	MRA	Minimum Reception Altitude.
FL	Flight Level.	MSA	Minimum Safe Altitude or Minimum Sector Altitude.
FMS	Flight Management System.		
GADO	General Aviation District Office.	MSL	Mean Sea Level.
GMT	Greenwich Mean Time.	NAS	National Airspace System.
GNC	Global Navigational Chart.	NAT OTS	North Atlantic Organized Track System.
GPH	Gallons Per Hour.		
GPS	Global Positioning System.	NDB	Non-Directional Beacon.
GRI	Group Repetition Interval.	NM	Nautical Mile or Miles.
GS	Ground speed or Glide Slope.	NMPH	Nautical Miles Per Hour.
GSPD	Ground speed.	NMS	Navigation Management System.
HDG	Heading.	NoPT	No Procedure Turn.
HF	High Frequency.	NOS	National Oceanic Service.
HSI	Horizontal Situation Indicator.		
IAS	Indicated Airspeed.	OBS	Omni Bearing Selector.
ICAO	International Civil Aviation Organization.	ONC	Operational Navigational Chart.
IFIM	*International Flight Information Manual.*	P	Precision.
		PCA	Polar Cap Anomaly.
IFR	Instrument Flight Rules.	PNR	Point-of-No-Return.
ILS	Instrument Landing System.	PPH	Pounds Per Hour.
IMC	Instrument Meteorological Conditions.	PT	Procedure Turn.
INS	Inertial Navigational System.	RB	Relative Bearing.
IRS	Inertial Reference System.	REM	Remaining.
JNC	Jet Navigational Chart.	RETA	Revised Estimated Time of Arrival.
KTS	Knots.	RGE	Range.
LCC	LORAN-C Navigational Chart.	RMI	Radio Magnetic Indicator.
LDA	Localizer-type Directional Aid.	RNAV	Area Navigation.
LED	Light Emitting Diodes.	RPU	Receiver/Processor Unit.
L/MF	Low/Medium Frequency.	SDF	Simplified Directional Facility.
LOC	Localizer	SID	Standard Instrument Departure or Sudden Ionospheric Disturbance.
LOP	Line of Position.		
LORAN	Long Range Navigation	SNR	Signal-to-Noise Ratio.
MAA	Maximum Authorized Altitude.	STAR	Standard Terminal Arrival.
MAP	Missed Approach Point.		
MB	Magnetic Bearing.	TACAN	Tactical Air Navigation.
MDA	Minimum Descent Altitude.	TAS	True Airspeed.
MEA	Minimum En route Altitude.	TC	True Course.
MFD	Multifunction Display.	TH	True Heading.
MH	Magnetic Heading.	TKE	Track Angle Error.
MLS	Microwave Landing System.	TRK	Track.
		TTG	Time-To-Go.

UHF	Ultra High Frequency.	VOR	Very High Frequency Omnidirectional Range.
UTC	Coordinated Universal Time.		
VAR	Variation.	VOT	VOR Test.
VFR	Visual Flight Rules.	WAC	World Aeronautical Chart.
VHF	Very High Frequency.	WCA	Wind Correction Angle.
VLF	Very Low Frequency.	WX	Weather.
VMC	Visual Meteorological Conditions.		
VNAV	Vertical Navigation.	XTD	Cross Track Distance.

Glossary

NAUTICAL 0 10 20 30 40 50 60 70 80

absolute accuracy—The ability to determine true geographic position.

accelerometers—Sensitive devices used in inertial navigation systems to detect changes in aircraft velocity.

additional secondary factor—Calibration factors designed to compensate for the difference in propagation characteristics between land and sea.

air data system—A sophisticated system for measuring air temperature, pitot pressure, and static pressure. Used to obtain highly accurate TAS, Mach, altitude, and static air temperature. Also called an "air data computer."

Aircraft Approach Category—A grouping of aircraft based on a speed of 1.3 times the stall speed in the landing configuration at maximum gross landing weight.

airway—A control area or portion thereof in the form of a corridor, the centerline of which is defined by radio navigational aids.

area navigation—Navigation along random routes within the area of coverage of referenced facilities, or within the limits of self-contained area navigational aids, precluding the need to overfly specific navigational facilities.

azimuth—Angle or direction.

baseline—The imaginary great circle line connecting a LORAN-C Master station with a Secondary station, or the imaginary great circle line connecting two OMEGA or two VLF stations.

baseline extension—The extension of the baseline beyond the LORAN-C Master or Secondary station, or beyond an OMEGA or VLF station.

bearing—The direction to or from any one point to any other, normally expressed either in terms of True or Magnetic North, or in terms of degrees from 0 to 360 relative to the aircraft heading.

calibrated airspeed—Indicated airspeed corrected for installation error.

changeover point—A point along a route or airway segment between two adjacent navigational facilities or waypoints where navigational guidance changes from one facility to the next.

coast-in—The first point of land reached when flying from seaward.

coast-out—The last point of land crossed when departing to seaward.

co-location—The pairing of related navigational facilities for the purpose of providing course and distance information from a common point; normally, DME (TACAN) is paired with VOR, ILS, or LOC.

compass heading—Magnetic heading corrected for magnetic deviation.

Compass Locator—A low power L/MF band radio beacon normally installed at the site of the outer or middle marker of an instrument landing system. It can be used for navigation at distances of approximately 15 miles, or as authorized in the approach procedure.

compulsory reporting point—A geographical location in relation to which the position of the aircraft must be reported to ATC; depicted on aeronautical charts by a solid triangle.

course—The intended direction of flight, measured normally either by reference to True or Magnetic North (but sometimes by reference to Grid North).

course line computer—A cockpit device to determine magnetic bearing and distance from the aircraft to a designated waypoint. Commonly called an ''RNAV.''

cross track distance—The distance and direction right or left between the aircraft and the desired track.

dead reckoning—The directing of an aircraft and the determination of its position by the application of direction and speed information to a previous position.

desired track—In VOR/DME based RNAV, the Rhumb Line course between selected waypoints. In long-range navigation, the Great Circle course between selected waypoints.

deviation—Compass error caused by uncorrectable magnetism within the aircraft.

differential GPS—An addition to the NAVSTAR/GPS system which uses corrections from a ground station for the purpose of enhancing NAVSTAR/GPS accuracy for all users within range of the associated ground station.

direct wave—Radiated energy which follows a line-of-sight path between the source and the receiver or sensor.

directional gyro—A gyroscopically stabilized heading indicator.

diurnal effect—Variation in the height of the ionosphere caused by sunlight; a daily phenomenon.

drift—(1) Displacement from the direction in which the aircraft is headed caused by a crosswind component. (2) The difference between reported position and true position.

drift angle—The angle between aircraft heading and the desired track, expressed in terms of right or left according to the direction in which the aircraft has drifted.

elevation—Height above Mean Sea Level.

ephemeris—The position of an astronomical body.

extended range modal interference—Modal interference affecting OMEGA/VLF signal propagation at ranges that vary from 1,000 to 2,000 nautical miles.

flight level—A flight altitude flown at a fixed altimeter setting of 29.92 inches of mercury.

glide path—That portion of the glide slope that intersects the localizer course.

glide slope—Vertical guidance provided the aircraft during approach and landing.

great circle—The circle formed by a plane that passes through the center of a sphere.

ground wave—Radiated energy which follows the surface of the earth.

group repetition interval—The amount of time, in microseconds, between the start and stop of each LORAN-C Master station transmission. Commonly used for station identification.

gyro—Short for gyroscope. A rapidly spinning mass which uses the properties of inertia to maintain its alignment along a given plane.

heading—The direction in which the longitudinal axis of the aircraft is aligned, measured clockwise from a given reference.

Horizontal Situation Indicator—An instrument which combines both directional gyro information and a VOR/ILS course deviation indicator to provide a symbolic representation of the aircraft in relation to the course.

Instrument Landing System—A precision instrument approach system that provides both horizontal and vertical guidance.

isogonic line—A line of constant magnetic variation.

Jet Route—A network of air routes based on radials from VOR navigational facilities extending from 18000 feet MSL up to and including Flight Level 450.

lane ambiguity resolution—In OMEGA systems, the process of identifying the correct lane, or area between hyperbolic lines of position.

latitude—Angular distance measured north or south of the Equator from 0 to 90.

line of position—A line containing all possible geographic positions of an observer at a given instant of time.

longitude—Any Great Circle which passes through Earth's poles. Measured in angular distance east or west from the zero meridian.

magnetic bearing—The bearing, measured clockwise using magnetic north as a reference datum, to or from an object or navigational facility.

magnetic north—The earth's magnetic north pole, located at approximately latitude 73 North, longitude 110 West.

magnetic variation—The difference in degrees east or west that magnetic north varies from true north for any given location.

Maximum Authorized Altitude—The published altitude representing the maximum usable altitude or flight level. On a Federal airway, the highest altitude at which reliable reception of navigational signals is assured.

meridian—A line of longitude.

Minimum Descent Altitude—The lowest altitude above Mean Sea Level to which descent is authorized on final approach or during circle-to-land maneuvering in execution of a standard instrument approach procedure where no electronic glide slope is provided.

Minimum En Route Altitude—The lowest published altitude between radio fixes which assures acceptable navigational signal coverage and meets obstacle clearance requirements along and for the full width of the route between those fixes.

Minimum Obstruction Clearance Altitude—The lowest published altitude between radio fixes which meets obstacle clearance requirements along and for the full width of the route segment between fixes and which assures acceptable navigational signal coverage within 22 nautical miles of the associated facility.

Minimum Sector Altitude—Altitudes depicted on approach charts providing at least 1,000 feet of obstacle clearance within a 25-mile radius of the primary approach facility. Emergency use only; acceptable navigational coverage is not assured.

modal interference—Errors resulting from the inability of the receiver to differentiate between ground and sky waves.

night effect—An overlapping of ground and sky waves, common at sunrise and sunset, resulting in temporarily erratic indications.

non-directional beacon—A radio beacon transmitting signals in all directions that can be used for homing, tracking, and orientation.

non-precision approach—Any instrument approach in which electronic glide slope information is not provided.

on request reporting point—A geographic location in relation to which aircraft position can be reported. Depicted on aeronautical charts by an open triangle.

orthogonal—Mutually perpendicular. The primary aircraft axes are orthogonally related.

parallels—Lines of latitude. Small Circles parallel to the equator.

pilotage—Navigation by reference to terrain features, both natural and man-made, usually with the aid of an appropriate aeronautical chart.

polar cap anomaly—In OMEGA/VLF systems, a propagation anomaly affecting signals traversing polar regions; caused by solar flares.

precision approach—Any instrument approach in which electronic glide slope information is provided.

procedure turn—A maneuver designed to reverse direction and establish the aircraft on the intermediate approach segment or final approach course that is part of a standard instrument approach procedure.

propagation anomalies—Inconsistencies in signal phase or frequency.

pseudo range—Distance determined by an uncorrected time measurement.

radar—An acronym for Radio Range and Detecting. A radio detection device, utilizing pulses of microwave length energy, to provide range, azimuth, and/or elevation to objects.

radar transponder—See "transponder."

radial—A magnetic bearing extending from a VOR, VORTAC, or TACAN navigation facility.

Radio Magnetic Indicator—An aircraft navigational instrument that combines a gyro compass with one or more directional indicators for the purpose of displaying magnetic direction to and from selected navigational facilities.

range—Distance.

rate aiding—The process of correcting LOP's taken at different times to equivalent times; requires heading and TAS.

relative bearing—The bearing, measured clockwise using the aircraft heading as the reference datum, to an object or navigational facility.

repeatable accuracy—The ability to return to a previous position, normally measured in nautical miles.

rho—Navigational term for distance.

rhumb line—A line which makes the same angle with each meridian it crosses.

secondary radar system—A radar system that can interrogate transponders in order to display their coded responses.

shore line effect—A refraction or bending of the ground wave caused by the difference in conductivity between land and water.

skip zone—The gap in radio wave transmission that exists between the outer limits of ground wave reception and the beginning of sky wave reception.

sky wave—Radiated energy which has been refracted by an ionized layer of the atmosphere back toward the surface of the earth.

slant range—The straight line distance between the aircraft and the DME range facility.

slant range error—The difference between slant range and range over the ground.

Standard Service Volume—Reception limits expressed in nautical miles, by altitude, of unrestricted navigational facilities for random and unpublished (direct) route navigation.

sudden ionospheric disturbance—Change in the height of the ionosphere, normally caused by solar flares.

Tactical Air Navigation—A UHF electronic rho-theta air navigation aid which provides suitably equipped aircraft (primarily military) with a continuous indication of bearing and distance to the TACAN facility.

terminal routing—A direct route from an initial approach fix to the inbound approach course where a procedure turn is neither required nor allowed without specific clearance.

theta—Navigational term for azimuth.

time bias—An error caused by differences between time standards.

torquing—A process of applying force to a gyro to intentionally induce precession (drift) so as to correct for the effects of earth rotation and aircraft movement.

track—The actual course made good by an aircraft.

track angle error—The difference, in degrees and direction right or left, between the actual aircraft track and the desired track.

tracking—The process of adjusting aircraft heading to compensate for the effects of wind drift.

transponder—A receiver/transmitter that will generate a reply signal on one frequency, after being properly interrogated on another. If the interrogation comes from a ground-based radar facility, the reply will be displayed on that facility's radar screens.

triple mixing—A process of statistically integrating the outputs from three separate inertial navigational systems for increased overall accuracy.

true north—One end of the axis about which the earth rotates.

vector—Navigational guidance in the form of specific headings, based on the use of radar.

Very High Frequency Omnidirectional Range—A ground-based electronic navigation aid transmitting in the very high frequency range providing 360 degrees of azimuth information. The basis of the present U.S. National Airspace System.

Victor Airways—A network of air routes based on radials from VOR navigational facilities extending from a minimum of 1200 feet above ground level, up to but not including 18000 feet MSL.

VORTAC—A navigational aid providing VOR azimuth, TACAN azimuth, and TACAN DME at one site. A combined VOR/TACAN facility.

voting—A process of comparing the outputs of three systems for the purpose of identifying and eliminating the least correct system.

waypoint—Any geographical fix defined either in terms of radial and distance from a VOR facility, or in terms of latitude and longitude. Used in area navigation to describe route segments.

wind component (or factor)—That part of the wind vector that is aligned with the aircraft track; tailwind components are positive factors, increasing ground speed over airspeed, and headwind components are negative factors, decreasing ground speed.

wind correction angle—The difference, in degrees, between aircraft heading and course. Also called a "drift correction", and sometimes called a "crab angle", or "cut."

wrong way propagation—In OMEGA/VLF systems, an error caused by the receipt of a relatively strong OMEGA/VLF signal from the opposite direction of that expected.

Index